BIOCHEMISTRY RESEARCH TRENDS

KAEMPFEROL

BIOSYNTHESIS, FOOD SOURCES AND THERAPEUTIC USES

BIOCHEMISTRY RESEARCH TRENDS

Additional books in this series can be found on Nova's website under the Series tab.

Additional e-books in this series can be found on Nova's website under the eBooks tab.

BIOCHEMISTRY RESEARCH TRENDS

KAEMPFEROL

BIOSYNTHESIS, FOOD SOURCES AND THERAPEUTIC USES

TERESA GARDE-CERDÁN
AND
ANA GONZALO-DIAGO
EDITORS

nova publishers
New York

NOTICE TO THE READER

Additional color graphics may be available in the e-book version of this book.

Library of Congress Cataloging-in-Publication Data

Names: Garde-Cerdan, Teresa, editor. | Gonzalo-Diago, Ana, editor.
Title: Kaempferol : biosynthesis, food sources and therapeutic uses / editors, Teresa Garde-Cerdan and Ana Gonzalo-Diago (Seccibon de Viticultura y Enologbia, Servicio de Investigacibon y Desarrollo Tecnolbogico Agroalimentario (CIDA) Instituto de Ciencias de la Vid y del Vino (ICVV), Logrodno, Spain, and others).
Description: Hauppauge, New York : Nova Science Publisher's, Inc., [2016] | Series: Biochemistry research trends | Includes index.
Identifiers: LCCN 2016030877 (print) | LCCN 2016031716 (ebook) | ISBN 9781634858281 (hardcover) | ISBN 9781634858588 (ebook) | ISBN 9781634858588 ()
Subjects: LCSH: Flavonoids.
Classification: LCC QK898.F5 K33 2016 (print) | LCC QK898.F5 (ebook) | DDC 575.4/359--dc23
LC record available at https://lccn.loc.gov/2016030877

Published by Nova Science Publishers, Inc. † New York

CONTENTS

Preface **vii**

Chapter 1 Kaempferol in Grape and Wine **1**
Riccardo Flamini, Mirko De Rosso and Annarita Panighel

Chapter 2 Flavonols: Enhancement by Using Elicitors **19**
Rocío Gil-Muñoz

Chapter 3 Therapeutic Uses of Kaempferol: Anticancer and Antiinflammatory Activity **71**
Ignacio Gutiérrez-del-Río, Claudio J. Villar and Felipe Lombó

Chapter 4 Kaempferol: Review on Natural Sources and Bioavailability **101**
Muhammad Jahangir Hossen, Md Bashir Uddin, Syed Sayeem Uddin Ahmed, Zhi-Ling Yu and Jae Youl Cho

Chapter 5 Kaempferol Glycosides in *Crocus*: Sources, Biosynthesis, and Uses **151**
Natalia Moratalla-López, Cándida Lorenzo, Gonzalo L. Alonso and Ana M. Sánchez

Chapter 6 Saffron Crocus (*Crocus sativus* L.) as a Source of Kaempferol **197**
Keti Zeka and Randolph Arroo

Editors' Contact Information **217**

Index **219**

PREFACE

Kaempferol is a natural plant product known for its health promoting effect and study for its pharmacological and nutraceutical activities. It is common in vegetables, fruits, plants and herbal medicines. Studies have shown that it reduces cancer, arteriosclerosis, cardiovascular disorders, and serve as antioxidant and anti inflammatory. For it, this book presents an overview of the biosynthesis, food sources and therapeutic uses of this promising compound.

In Chapter 1, Flamini et al. summarize the biosynthesis of kaempferol derivatives and the other flavonols identified in grape. Also, the main analytical methods used for the characterization of kaempferol derivatives, in particular high-resolution mass spectrometry, focused on the different glycoside forms identified are shown. In addition, these authors briefly sum up the nutraceutical characteristics of flavonols with particular emphasis for kaempferol, by reporting the contents of kaempferol derivatives in grape and wine.

Many researchers have reported different experiences about the use of different abiotic and biotic elicitors in several cultivars obtaining contradictory results according to the compounds used, the doses and the application moment. The influence of different elicitors in flavonols composition and more specifically in kaempferol is presented in Chapter 2. Gil Muñoz, makes an overview of the use of elicitation as a strategy to protect plants against pests and diseases and also to enhance the secondary metabolite concentration in plants and cell cultures. Phenolic compounds can be found among these secondary metabolites; being a very spread group with comprise flavonol compounds, and among the flavonols, kaempferol is a very important compound due to its application in medicine, and consequently in human health. Different elicitors can be used to obtain an enhancement of flavonols, and consequently of kaempferol in different crop plants. Thus, in this chapter, a general vision about the use of elicitors and the influence of them in the concentration of phenolic compounds and specifically in flavonols is shown.

In Chapter 3, Gutiérrez-del-Río et al. show the antitumor and the antiinflammatory activities of kaempferol on diverse diseases. Furthermore, these authors highlighted the importance of flavonoids in co-therapy. Strong epidemiological evidences show an inverse relationship between frequent consumption of fruits and vegetables and the incidence of some types of health disorders, including certain cancers like colon, lung, ovarian, gastric or pancreatic cancer. Recent investigations have identified several diet phytochemicals that could be involved in cancer prevention, as for example flavonols like kaempferol. In spite of recent progress in cancer treatment, classic chemotherapy has got numerous limitations like a low selectivity and tumor recurrence due to resistance development. These circumstances make necessary the search of new chemopreventive or chemotherapeutic agents. Kaempferol is gaining importance as phytochemical because of its wide range of pharmacological properties. The anticancer effect of this flavonoid is mediated through different modes of action including antiproliferative activity, apoptosis induction, cell-cycle arrest, antiangiogenic and antimetastasic activities. However, kaempferol efficacy against *in vivo* tumor development and progression is impaired by its low bioavailability. One solution to this problem is the use of nanopharmaceuticals. Phytochemicals encapsulation in nanocarriers can improve their bioavailability, enhancing its specific cellular internalization by tumor cells, reducing phytochemical doses, and allowing the development of combined therapies including phytochemicals together with classical chemotherapy compounds.

Recently, researcher's interest has been focused on dietary products due to the several healthly benefits. Kaempferol is a flavonoid linked to diverse glycoside moieties and extremely plentiful in most edible plants such as tea, fruits and vegetables including *Allium cepa* (onion), *Camellia sinensis* (tea), *Citrus paradisi* (grapefruit), *Fragaria vesca* (strawberry), *Lactuca sativa* (lettuce), and *Morinda citrifolia* (Indian mulberry) as well as in medicinal plants like as *Acacia nilotica* (L), *Aloe vera* (L.), *Crocus sativus* (L.), *Euphorbia pekinensis* Rupr., *Ginkgo biloba* (L.), *Hypericum perforatum* (L.), *Phyllanthus acidus* (L.), *Ribes nigrum* (L.), *Rosmarinus officinalis* (L.), *Cerbera manghas*, and *Persicaria chinensis* (L.) etc. Also, molecular mechanistic studies report that kaempferol modulates a number of key elements in cellular signal transduction pathway linked to apoptosis, angiogenesis, inflammation, and metastasis. Chapter 4 provides several natural sources of kaempferol with its pharmacokinetics (oral availability) and safety. Jahangir Hossen et al. show the dietary wealth of kaempferol with its bioavailability and emphasize the potential possibilities of kaempferol as a prospective novel candidate for future drug development.

Chapter 5 and Chapter 6 illustrate the use of Crocus sativus L. as a specie that stands out for the highly valued stigmas of its flowers used as spice, food additive and medicinal drug. The genus Crocus (Iridaceae) comprises about 160 species occurring in the wild in Europe, the Middle East and North Africa and grown as ornamentals all over the world for their white, yellow, pale-brown, purple to lilac, mauve and blue flowers. Among these species, the cultivated saffron (Crocus sativus L.) stands out for the highly valued stigmas of its flowers used as spice, food additive and medicinal drug. Kaempferol glycosides are major flavonoids in the flowers of this genus (70-90% of the total content in the perianth) and are also present in the leaves. In Chapter 5, Moratalla-López et al. show the occurrence of kaempferol and its glycosidic patterns in different Crocus species. Biosynthesis aspects such as pathways for sequential glycosylation of kaempferol together with the role of kaempferol in the current food, therapeutic or ornamental uses and in other potential uses of the species of the genus Crocus are also analysed. These authors pay special to saffron (Crocus sativus L.) and the growing interest of its tepals, which are rich in kaempferol 3-O-sophoroside, as a source of antioxidants and active principles. In Chapter 6, Zeka and Arroo study the medicinal uses of saffron and the ways of purification, isolation and identification because rather than a waste product of the saffron spice production, the petals can be a readily exploitable good source of kaempferol for many applications. Moreover, the different mechanisms of actions of kaempferol are shown.

In: Kaempferol ISBN: 978-1-63485-828-1
Editors: T.Garde-Cerdán and A.G. Diago ©2016 Nova Science Publishers, Inc.

Chapter 1

KAEMPFEROL IN GRAPE AND WINE

Riccardo Flamini*, Mirko De Rosso and Annarita Panighel

Council for Agricultural Research and Economics (CREA) – Viticulture and Oenology, Conegliano, Treviso, Italy

ABSTRACT

The first paragraph summarizes the biosynthesis of kaempferol derivatives and the other flavonols identified in grape; the second presents the main analytical methods used for the characterization of kaempferol derivatives, in particular high-resolution mass spectrometry, and focalizes on the different glycoside forms identified. The last paragraph summarizes the nutraceutical characteristics of flavonols with particular emphasis for kaempferol, by reporting the contents of kaempferol derivatives in grape and wine.

1. BIOSYNTHESIS OF FLAVONOLS AND KAEMPFEROL DERIVATIVES IN GRAPE

Flavonols are secondary metabolites distributed in a wide range of vegetable food sources, in higher plants they predominantly accumulate as glycosides.

* Corresponding Author address; Email: riccardo.flamini@crea.gov.it.

Among various fruits the grape is particularly rich in phenolic compounds, in particular flavonols, flavanols and anthocyanins (Macheix et al., 1990; Downey et al., 2006). Flavonoids are mainly located in the outer epidermis of the berry skin since they act as UV-protecting agents by absorption of sunlight UV-A and UV-B. Their synthesis begins in the flower buttons, being its highest concentrations found few weeks after veraison, then the content stabilizes during the early fruit development and it decreases as the size of the berry increases (Downey et al., 2003; Hermosín-Gutiérrez et al., 2012).

Molecular structure of flavonols consists of an aromatic ring A condensed to heterocyclic ring C and attached to a second aromatic ring B. Their antioxidant activity is due to the presence of phenolic hydroxyl groups attached to the aromatic rings. Principal flavonols in grape are quercetin (Q), kaempferol (Kf), and myricetin (Mr) present as glycosides. In general, the sugar moiety is conjugated at the 3-position in the C ring and no C-glycosilated compounds have been found. Monoglycosides are mainly glucose and glucuronic acid 3-*O*-glycosides, while galactose, rutinose or pentose derivatives were found in minor quantities (Mattivi et al., 2006; Makris et al., 2006; Castillo-Muñoz et al., 2009; Azuma et al., 2012; Koyama et al., 2012). Other grape flavonols are isorhamnetin (Is), laricitrin (Lr), and syringetin (Sy) (Cheynier and Rigaud, 1986; Mattivi et al., 2006) (Figure 1).

Biosynthesis and content of flavonols depend on the genotype and the flavonol profile is studied as important chemo-taxonomic parameter, even if in some extent it can be modulated by biotic and abiotic factors (Mattivi et al., 2006; Castillo-Muñoz et al., 2007; Figueiredo-González et al., 2012; Liang et al., 2012a; 2012b). White and light red grape varieties synthesize principally the B-ring mono and di-substitutes molecules, such as kaempferol, quercetin, and isorhamnetin, while red grapes also accumulate the tri-substitute compounds myricetin, laricitrin and syringetin (Mattivi et al., 2006; Castillo-Muñoz et al., 2010). The biosynthesis also differs by the type of substitution in B-ring: kaempferol is monohydroxylated in 3' position, quercetin is dihydroxylated in positions 3' and 4', myricetin trihydroxylated in positions 3', 4' and 5', isorhamnetin is a methylated quercetin, laricitrin and syringetin are two methylated forms of myricetin (structures in Figure 1). Quercetin is the major flavonol in white grapes (e.g., Chardonnay, Riesling, Viogner and Sauvignon Blanc it represents over 70% of total flavonols) and in some light red/rosè varieties as Nebbiolo, Pinot Noir, Sangiovese and Gewürztraminer, whereas myricetin is the major flavonol in most of the red varieties, such as Cabernet Sauvignon, Sagrantino and Teroldego (Mattivi et al., 2006). In general, the methylated derivatives isorhamnetin, laricitrin, and syringetin are present in smaller amounts. Because of the copigmentation with anthocyanins, flavonols play also an important role in color stability of red wines (Flamini et al., 2013 and references cited herein).

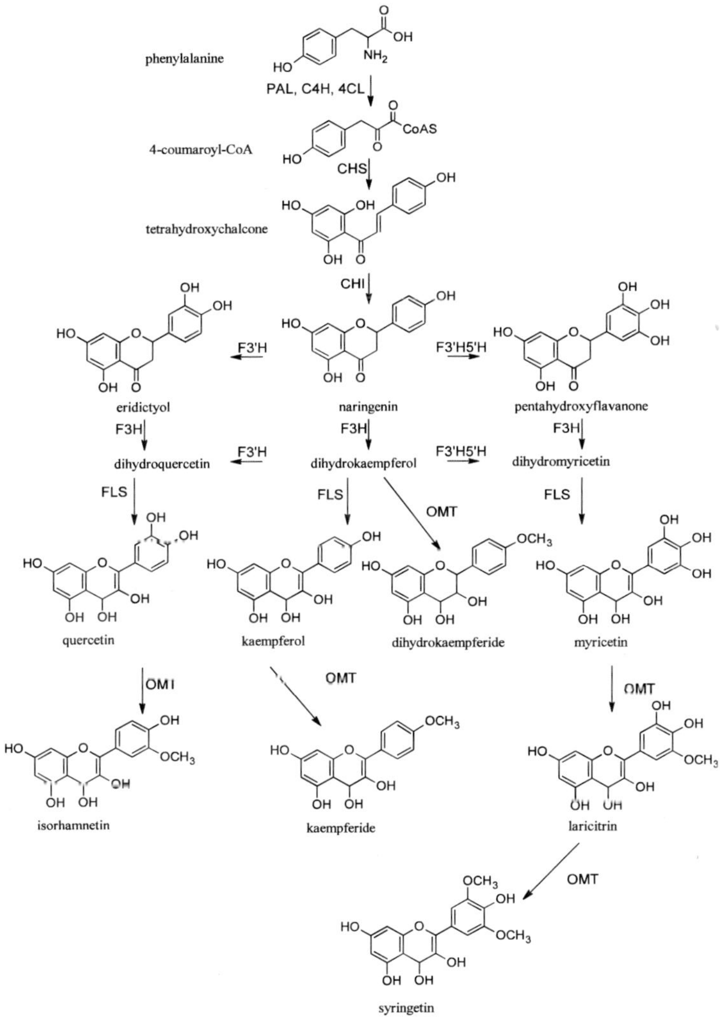

Figure 1. Putative biosynthetic pathway of flavonols in grape. The trihydroxylated flavonols myricetin, laricitrin and syringetin are lacking in the berries of white grapes. PAL, phenylalanine ammonia lyase; C4H, cinnamate-4-hydroxylase; 4CL, 4-coumaroyl CoA-ligase; CHS, chalcone synthase; CHI, chalcone isomerase; F3'H, flavonoid-3'-hydroxylase; F3'5'H, flavonoid-3,'5'-hydroxylase; F3H, flavanone-3-hydroxylase; FLS, flavonol synthase; OMT, *O*-methyltransferase.

Recently, also diglycoside flavonols, such as myricetin hexoside-glucuronide, myricetin *O*-di-hexoside, quercetin *O*-di-hexoside, syringetin *O*-di-hexoside, isorhamnetin rutinoside and kaempferol rutinoside, were found in grape (De Rosso et al., 2014; Vrhovsek et al., 2012).

In general, flavonols are not acylated at the sugar moiety but recently three *p*-coumaroyl hexoside derivatives were putatively identified in grape: a dihydrokaempferide-3-*O*-*p*-coumaroylhexoside-like flavanone, isorhamnetin-3-*O*-*p*-coumaroylglucoside, a kaempferide-*p*-coumaroylhexoside-like flavone (Panighel et al., 2015). Their structures are shown in Figure 2. List of the flavonols identified in grape, is reported in Table 1.

Pathway in Figure 1 shows as the chemical structure of the flavonoids is strictly related to their biosynthesis. They are produced by the phenylpropanoids pathway which converts phenylalanine into 4-coumaroyl-CoA and later tetrahydroxychalcone. Flavonols are primarily synthesized from dihydroflavonols via the activity of flavonol synthase (FLS), a non-heme, 2-oxoglutarate-dependent oxygenase, although there is evidence that suggests that also anthocyanidin synthase (ANS) can generate flavonols via flavan-3,4-diol intermediates (Turnbull et al., 2004; Owens et al., 2008). Four dihydroflavonols found in grape (dihydroquercetin-3-rhamnoside, dihydroquercetin-3-galactoside, dihydroquercetin-3-glucoside, and dihydrokaempferol-3-rhamnoside) would confirm this pattern (Trousdale et al., 1983; Souquet et al., 2000; Masa et al., 2007; De Rosso et al., 2015).

Synthesis of kampferide and dihydrokampferide is potentially linked to a grape *O*-methyltransferase activity, so we can assume that a promoted activity - e.g., by a pathogen stress (Gichi, 2011) - might induce higher levels of these compounds in grape.

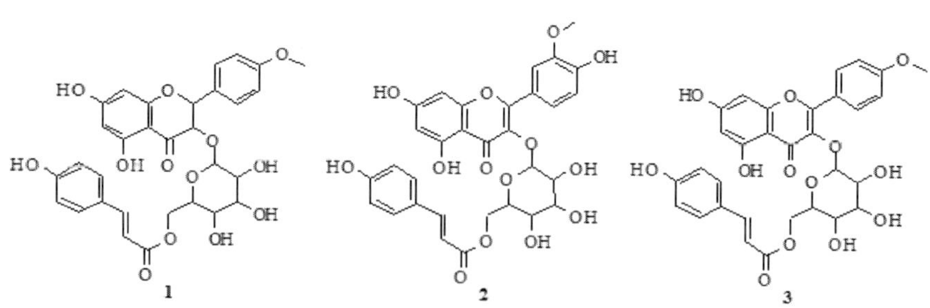

Figure 2. Putative *p*-coumaroyl hexoside flavonols identified in grape: 1. dihydrokaempferide-3-*O*-*p*-coumaroylhexoside; 2. isorhamnetin-3-*O*-*p*-coumaroylglucoside; 3. kaempferide-3-*O*-*p*-coumaroylhexoside (Panighel et al., 2015).

Table 1. Flavonols and dihydroflavonols identified in grape

	Formula	theoretical $[M\text{-}H]^-$ (*m/z*)
Flavonols		
Myricetin 3-*O*-glucuronide	$C_{21}H_{18}O_{14}$	493.0624
Myricetin 3-*O*-galactoside	$C_{21}H_{20}O_{13}$	479.0831
Myricetin hexoside-glucuronide	$C_{27}H_{28}O_{19}$	655.1152
Myricetin 3-*O*-glucoside	$C_{21}H_{20}O_{13}$	479.0831
Myricetin *O*-di-hexoside	$C_{27}H_{30}O_{18}$	641.1359
Quercetin 3-rhamnosylglucoside (rutin)	$C_{27}H_{30}O_{16}$	609.1461
Quercetin 3-*O*-galactoside	$C_{21}H_{20}O_{12}$	463.0882
Quercetin 3-*O*-glucuronide	$C_{21}H_{18}O_{13}$	477.0675
Quercetin 3-glucosylgalactoside [a]	$C_{27}H_{30}O_{17}$	625.1410
Quercetin 3-glucosylxyloside [a]	$C_{26}H_{28}O_{16}$	595.1305
Quercetin 3-*O*-glucoside	$C_{21}H_{20}O_{12}$	463.0882
Quercetin 4'-*O*-glucoside [b]	$C_{21}H_{20}O_{12}$	463.0882
Quercetin 3-*O*-rhamnoside [b]	$C_{21}H_{20}O_{11}$	447.0933
Quercetin 3,4'-diglucoside [b]	$C_{27}H_{30}O_{17}$	625.1410
Laricitrin 3-*O*-hexoside	$C_{22}H_{22}O_{13}$	493.0988
Laricitrin 3-*O*-glucuronide	$C_{22}H_{20}O_{14}$	507.0780
Isorhamnetin 3-*O*-*p*-coumaroylglucoside [c]	$C_{31}H_{28}O_{14}$	623.1406
Isorhamnetin *O*-rhamnosyl-hexoside	$C_{28}H_{32}O_{16}$	623.1618
Isorhamnetin 3-*O*-glucuronide	$C_{22}H_{20}O_{13}$	491.0831
Isorhamnetin 3-*O*-hexoside	$C_{22}H_{22}O_{12}$	477.1038
Kaempferol 3-*O*-galactoside	$C_{21}H_{20}O_{11}$	447.0933
Kaempferol *O*-rhamnosyl-hexoside	$C_{27}H_{30}O_{15}$	593.1512
Kaempferol 3-*O*-glucuronide	$C_{21}H_{18}O_{12}$	461.0725
Kaempferol 3-*O*-glucoside	$C_{21}H_{20}O_{11}$	447.0933
Kaempferol 3-glucosylarabinoside [a]	$C_{26}H_{28}O_{15}$	579.1355
Kaempferide *p*-coumaroylhexoside [c]	$C_{31}H_{28}O_{13}$	607.1457
Syringetin 3-*O*-galactoside	$C_{23}H_{24}O_{13}$	507.1144
Syringetin 3-*O*-glucoside	$C_{23}H_{24}O_{13}$	507.1144
Syringetin *O*-di-hexoside	$C_{29}H_{34}O_{18}$	669.1672
Dihydroflavonols		
Dihydrokaempferide 3-*O*-*p*-coumaroylhexoside [c]	$C_{31}H_{30}O_{13}$	609.1614
Dihydroquercetin 3-*O*-rhamnoside	$C_{21}H_{22}O_{11}$	449.1089
Dihydroquercetin 3-*O*-glucoside [d]	$C_{21}H_{22}O_{12}$	465.1038
Dihydroquercetin 3-galactoside [d]	$C_{21}H_{22}O_{12}$	465.1038
Dihydroquercetin 3-*O*-hexoside	$C_{21}H_{22}O_{12}$	465.1038
Dihydrokaempferol 3-*O*-rhamnoside	$C_{21}H_{22}O_{10}$	433.1140

De Rosso et al., 2014; [a]Chenier and Rigaud, 1986; [b]Vrhovsek et al., 2012; [c]Panighel et al., 2015; [d]Masa et al., 2007.

2. Identification of Grape Kaempferol Derivatives by High-Resolution Mass Spectrometry

If no standard compounds are available, UV-Vis spectrophotometry and mass spectrometry (MS) are actually the more effective tools to achieve a first putative identification of the metabolites in complex matrix, such as grape extracts and wines (Flamini et al., 2015). Due to high polarity of polyphenols, these techniques are usually coupled to reverse-phase High Performance-Liquid Chromatography (HPLC) (Di Stefano et al., 2008; Flamini et al., 2008). In general, Liquid-Chromatography/Mass Spectrometry (LC/MS) and Multiple Mass Spectrometry (MS/MS and MSn) are the most effective tools for the structural characterization of low-molecular weight (MW) polar compounds, such as flavonols, which were studied by operating in both positive and negative ionization mode. Analysis of sequence elution from the chromatographic column of the compounds belonging to the same chemical class can provide further confirmation of the identification (Castillo-Muñoz et al., 2007; De Rosso et al., 2014).

Table 2. Mass spectrometry (MS) and UV-Vis data used for identification of the kaempferol derivatives in grape

compound	formula	UV-vis (nm)	[M-H]$^-$	[M+H]$^+$	MS/MS
			(theoretical *m/z*)		(*m/z*)
kaempferol-3-*O*-galactoside [a,b,c]	$C_{21}H_{20}O_{11}$	266, 292(sh), 320(sh), 348	447.0933	449.1078	284.0326, 255.0299
kaempferol-3-*O*-glucoside [a,b,c]	$C_{21}H_{20}O_{11}$	265, 298(sh), 320(sh), 348	447.0933	449.1078	284.0326, 255.0299
kaempferol-3-*O*-glucuronide [a,c,d]	$C_{21}H_{18}O_{12}$	265, 290(sh), 320(sh), 348	461.0725	463.0793	285.0405
kaempferol-3-*O*-rhamnoside [e,f]	$C_{21}H_{20}O_{10}$	264, 295(sh), 345	431.0984	433.1129	285.0405
kaempferol-rutinoside [a,e]	$C_{27}H_{30}O_{15}$	265, 295(sh), 347	593.1512	595.1657	285.0405, 255.0299
dihydrokaempferol-3-*O*-glucoside [e,f]	$C_{21}H_{22}O_{11}$		449.1089	451.1235	287.0561
kaempferol-3-*O*-glucosylarabinoside [g]	$C_{26}H_{28}O_{15}$	265, 298(sh), 349	579.1345	581.1501	
dihydrokaempferol-rhamnoside [a,b,e]	$C_{21}H_{22}O_{10}$	290, 340(sh)	433.114	435.1286	287.0561, 269.0455
kaempferide-*p*-coumaroylhexoside [h]	$C_{31}H_{28}O_{13}$		607.1457	609.1603	299.0561, 284.0326
dihydrokaempferide-*p*-coumaroylhexoside [h]	$C_{31}H_{30}O_{13}$		609.1614	611.1759	301.0718, 283.0619

UV-Vis, maxima in the spectrum (sh, shoulder); [M-H]$^-$ and [M+H]$^+$, accurate mass of pseudomolecular ion measured by high-resolution MS (negative and positive ionization mode, respectively); MS/MS, main fragments used for the identification of compound by multiple mass spectrometry ([a]De Rosso et al., 2014; [b]De Rosso et al., 2015; [c]Castillo-Muñoz et al., 2007; [d]Hilbert et al., 2015; [e]Masa et al., 2007; [f]Zhu et al., 2012; [g]Cheynier et al., 1986; [h]Panighel et al., 2015).

Kaempferol derivatives, such as 3-*O*-glucoside, 3-*O*-glucuronide, 3-*O*-galactoside, 3-*O*-rhamnoside, and rutinoside, have been currently identified in grape. Dihydrokaempferol-3-*O*-rhamnoside was found in Albariño, Lado, Loureiro, Treixadura, and Xerez grapes (Masa et al., 2007); recently, dihydrokaempferide-*p*-coumaroylhexoside and a kaempferide-*p*-coumaroylhexoside derivative were identified in Raboso Piave grape (Panighel et al., 2015). High-resolution MS and UV-Vis data which provided their identification are reported in Table 2; high-resolution MS/MS spectra are shown in Figures 3-9.

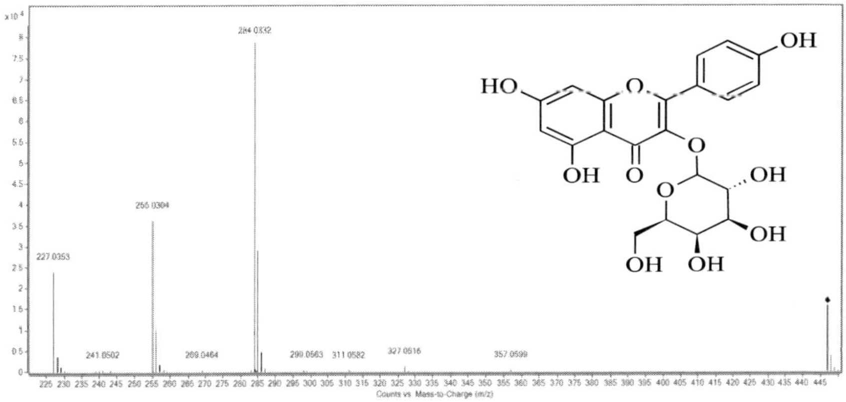

Figure 3. High-resolution multiple mass spectrometry (MS/MS) spectrum of kaempferol-3-O-galactoside.

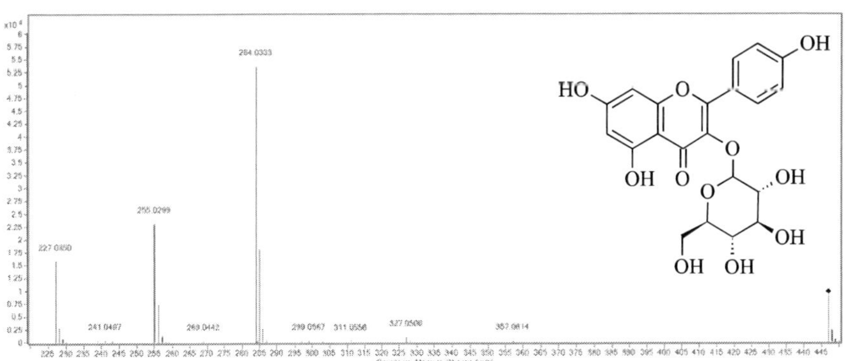

Figure 4. MS/MS spectrum of kaempferol-3-*O*-glucoside.

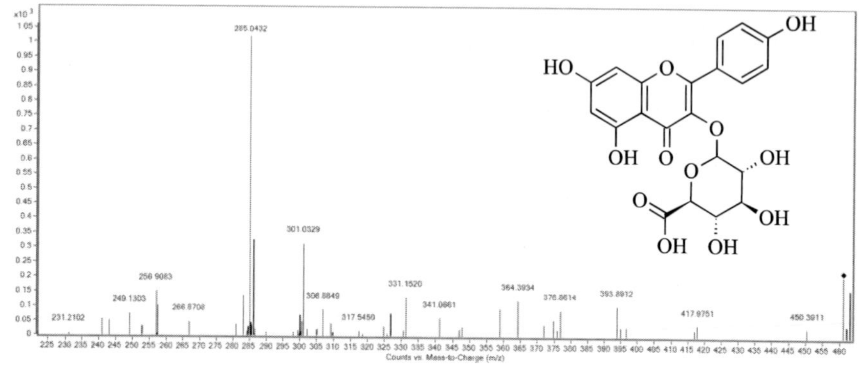

Figure 5. MS/MS spectrum of kaempferol-3-*O*-glucuronide.

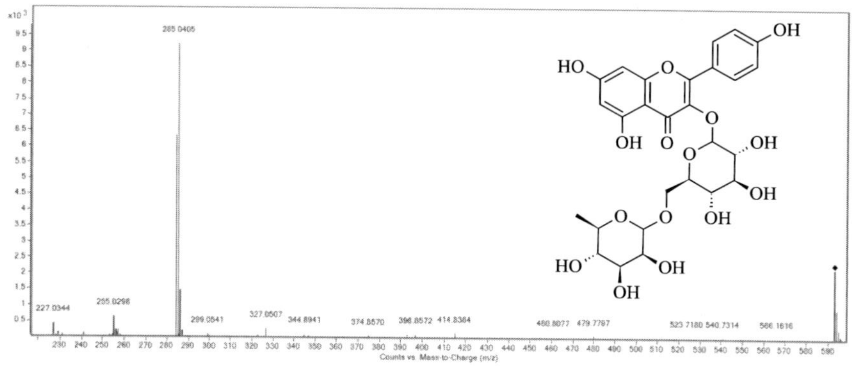

Figure 6. MS/MS spectrum of kaempferol-rutinoside.

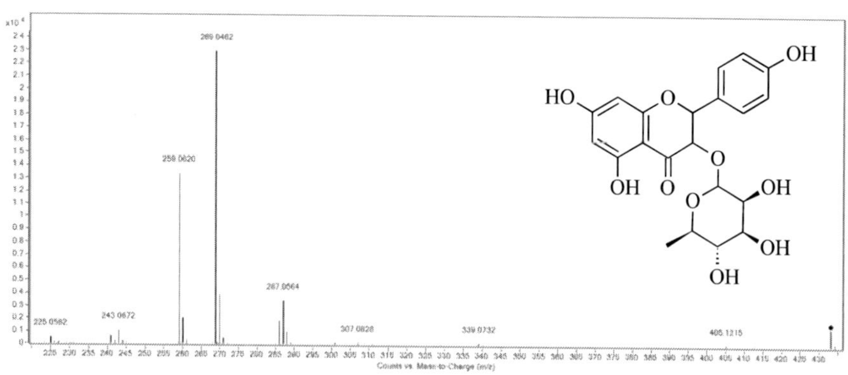

Figure 7. MS/MS spectrum of dihydrokaempferol-rhamnoside.

By application of low-medium collisional energy (20-30 eV) the MS/MS spectra of kaempferol galactoside and glucoside showed predominant homolytic cleavage and formation of the $[Y_0\text{-}H]^{-\bullet}$ ion at *m/z* 284.033 (Figures 3 and 4), whereas kaempferol 3-*O*-glucuronide and rutinoside showed higher formation of $[Y_0]^-$ ion *m/z* 285.040 (Figure 6) (De Rosso et al., 2014).

Putative identification of dihydrokaempferide-*p*-coumaroylhexoside mainly relied on the study of high-resolution MS/MS spectrum in Figure 8 (Panighel et al., 2015). Base peak of the spectrum is the signal at *m/z* 301.0721, corresponding to the aglycone ion $[Y_0]^-$ formed by heterolytic cleavage and *p*-coumaroylglucoside residue loss (-308 Da), the fragment at *m/z* 300.0645 is produced by homolytic cleavage and formation of the radical aglycone ion $[Y_0\text{-}H]^{-\bullet}$. Fragments at *m/z* 283.0619 and *m/z* 268.0379 were attributed to $[Y_0\text{-}H_2O]^-$ and $[Y_0\text{-}H_2O\text{-}CH_3]^{-\bullet}$ ions, respectively, those at *m/z* 163.0404, *m/z* 145.0297, *m/z* 119.0480 and *m/z* 307.0830 are correlated to *p*-coumaroylhexoside residue (data not showed) (Zhang et al., 2007; Vallverdú-Queralt et al., 2011; Abu-Reidah et al., 2013). The fragment at *m/z* 285.0405 matches that of the $[Y_0\text{-}H\text{-}CH_3]^-$ ion which is diagnostic of a OCH_3 group (Ablajan et al., 2013).

High-resolution MS/MS spectrum of putative kaempferide-3-*O*-*p*-coumaroylhexoside shows the signal at *m/z* 299.0565 as base peak, which potentially corresponds to the $[Y_0]^-$ ion, and an intense signal at *m/z* 298.0481 of $[Y_0\text{-}H]^{-\bullet}$ ion. The fragments at *m/z* 284.0337, *m/z* 283.0254, *m/z* 271.0581 and at *m/z* 240.0423 matched with the $[Y_0\text{-}CH_3]^{-\bullet}$, $[Y_0\text{-}H\text{-}CH_3]^-$, $[Y_0\text{-}CO_2]^-$ and $[Y_0\text{-}CH_3\text{-}CO_2]^{-\bullet}$ ions, respectively (Figure 9). Alternatively, chrysoeriol-7-*O*-*p*-coumaroylhexoside was proposed (Panighel et al., 2015).

Structures of two 3-*O*-*p*-coumaroylhexosides were confirmed by the absence of the $[M\text{-}H\text{-}H_2O]^-$ ion which indicates they are not *C*-glycoside but *O*-glycoside derivatives (Cuyckens et al., 2004). Also the study of the $[Y_0\text{-}H]^{-\bullet}/[Y_0]^-$ ratio was useful in determining the position of sugar moiety (Cuyckens et al., 2005). In particular, flavanone-7-*O*-glycosides not show formation of $[Y_0\text{-}H]^{-\bullet}$ ions, and 3-*O* and 7-*O*-glycoside flavonoids have different relative abundances of $[Y_0]^-$ and $[Y_0\text{-}H]^{-\bullet}$ ions, heterolytic cleavage being favored in the latter (Hvattum et al., 2003; Ablajan et al., 2006).

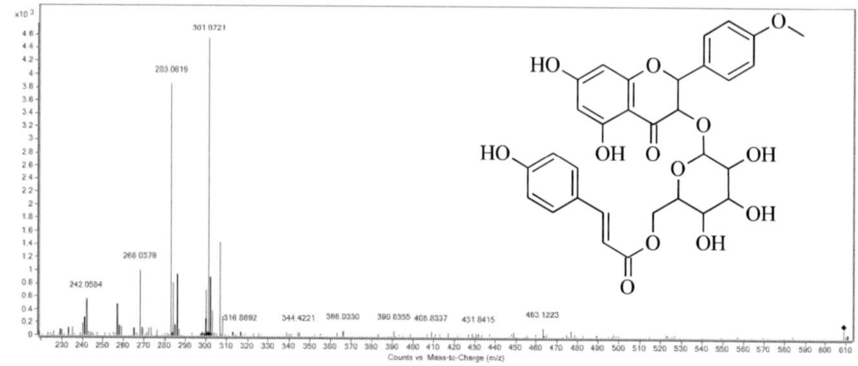

Figure 8. MS/MS spectrum of putative dihydrokaempferide-3-*O*-*p*-coumaroylhexoside-like compound.

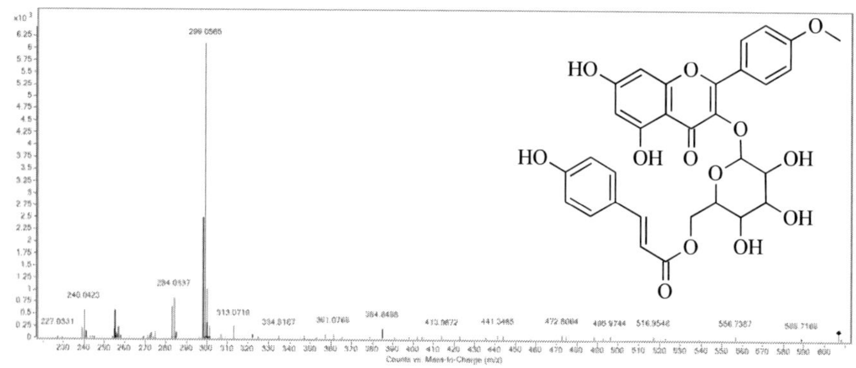

Figure 9. MS/MS spectrum of putative kaempferide-*p*-coumaroylhexoside flavone.

3. Kaempferol in Grape and Wine

In general, flavonols are characterized by relevant biological activities: act as potent inhibitors of LDL oxidation by protecting the cells against damage induced by reactive oxygen species; protect LDL against copper ion-induced oxidation; exhibit radical-scavenging activity and scavenge free radicals; and exhibit a general affinity to ATP-binding proteins linked to their structural analogy with ATP (Fuhrman et al., 2002; Tomás-Barberán et al., 2012).

Several in-vitro and in-vivo studies showed that kaempferol and some glycoside derivatives are characterized by antioxidant, anti-inflammatory, anticancer, and antimicrobial activities. In particular, kaempferol is a potent superoxide scavenger with IC_{50} of 0.5 μM, inhibits the activity of enzymes that create reactive oxygen species (ROS), can inhibit P450 enzymes and prevent the formation of carcinogenic promoters, has antiviral activity against several viruses, interferes with enzymes that are essential for the growth of some fungi, prevent lipid peroxidation (Calderón-Montano et al., 2011 and references cited herein).

The content of flavonols in grape can vary from 1 to 80 mg/Kg of fresh berry, mainly depending on the variety, the red cultivars being often richer than the white ones. Particularly high contents were found in some wild *Vitis* species, such as *V. palmate* (124 mg/Kg) and *V. riparia* (111 mg/Kg). Content is also influenced by the berry skin thickness: it was found that wines produced by thick skinned grapes (e.g., Cabernet Sauvignon) have higher amount of flavonols. Also agronomical and environmental factors can influence the amount (principally) and the profile (in secondary extent) of flavonols in grapes. In particular, their biosynthesis is greatly influenced by sunlight and, even if in less extent, it seems by the temperature. There is evidence that the flavonol biosynthesis is more strongly modulated by light with respect to other metabolites in the berry, in particular UV-B radiation activates the synthesis of flavonols while it is significantly decreased by shading. It was reported that day temperatures between 15-25°C followed by night temperatures 10-20°C produce grapes with higher amount in flavonols with respect to higher temperatures (30-35°C) (Flamini et al., 2013 and references cited herein).

Table 3. Contents of kaempferol derivatives found in grapes and wine

	Berry	Skin	Grape juice/Wine
	(mg/Kg)		(mg/L)
kaempferol	0.1-0.3 [a]		0-7.5 [a,b,c,d,e,f]
kaempferol-3-*O*-galactoside	0.01-0.93 [g]	0-4.2 [h]	
kaempferol rutinoside	2.8 [a]		0-2.08 [a,i]
kaempferol-3-*O*-glucuronide		0-9.3 [h]	
kaempferol-3-*O*-glucoside	0.03-26 [a,g,l]	0-17.9 [h]	1.3-12.0 [c,m]
kaempferol-3-*O*-glucosylarabinoside			2.9-9.1 [m]
kaempferide-*p*-coumaroylhexoside	0-0.05 [g] (as Kf-glucoside by LC/MS)		
dihydrokaempferide-*p*-coumaroylhexoside	0.1 [n] (as IS by LC/MS)		

[a]Gómez-Alonso et al., 2007; [b]Fang et al., 2007; [c]Ghiselli et al., 1998; [d]Padilla et al., 2005; [e]Rodríguez-Bernaldo de Quirós et al., 2009; [f]Stecher et al., 2001; [g]De Rosso et al., 2015; [h]Hilbert et al., 2015; [i]Hernández et al., 2006; [l]Montealegre et al., 2006; [m]Gutiérrez et al., 2005; [n]Panighel et al., 2015). IS, compound quantified on the internal standard signal.

Table 3 reports the contents of kaempferol derivatives found in *Vitis* grapes and wines. Kaempferol-3-*O*-glucoside is the major of these compounds in grape and, as a consequence, found in the wine, also kaempferol-3-*O*-glucuronide and kaempferol rutinoside can be present in relevant amounts. High contents of kaempferol-3-*O*-glucoside and kaempferol-rutinoside were reported in some port wine grapes, such as Touriga National, Tinta Amarela, Tinta Barocca, and Tinta Roriz (between 49-141 mg/Kg and 15-31 mg/Kg, respectively; Andrade et al., 2001).

Conclusion

Vitis grapes and wines have a flavonol profile often characterized by a number of compounds and ten different kaempferol derivatives were identified. Since their profile is mainly influenced by the cultivar, these compounds are important markers for chemotaxonomy of both red and white grape varieties.

Flavonols and kaempferol have been often associated to the quality of grapes and wines as important sources of antioxidants in the human diet. Because of their photo-protective role towards direct sunlight, the amounts in grapes, and as a consequence in wines, can be influenced by agronomical and environmental factors, such as canopy management, bunch exposure, irrigation, yield regulation, and timing of harvest (Downey et al., 2006 and references cited herein). As a consequence, the biosynthesis of these compounds can be favored by appropriate and targeted agronomical practices and environmental choices - made while meeting the oenological requirements of the products - by increasing the nutraceutical properties of grapes and wines.

The recent studies of metabolomics performed by high-resolution MS allowed to improve the knowledge of grape flavonols. The application of these analytical techniques in the next few years can induce further identification of new compounds - including Kf derivatives - and provide useful information on the metabolism of grape cells and tissues. For example, the identification of two Kf *p*-coumaroylglucoside derivatives suggests that also other acylated derivatives may be synthesized in grape. Probably, the knowledge of kaempferol chemistry in grape can be increased by characterization of Vitis grape varieties not still studied. Use of these analytical approaches is particularly promising for this aim.

REFERENCES

Ablajan K, Abliz Z, Shang X-Y, He J-M, Zhang R-P and Shi J-G (2006) Structural characterization of flavonol 3,7-di-O-glycosides and determination of the glycosylation position by using negative ion electrospray ionization tandem mass spectrometry. *Journal of Mass Spectrometry* 41:352–360.

Ablajan K and Tuoheti A (2013) Fragmentation characteristics and isomeric differentiation of flavonol O-rhamnosides using negative ion electrospray ionization tandem mass spectrometry. *Rapid Communications in Mass Spectrometry* 27:451–460.

Abu-Reidah IM, Arráez-Román D, Segura-Carretero A and Fernández-Gutiérrez A (2013) Extensive characterisation of bioactive phenolic constituents from globe artichoke (*Cynara scolymus L.*) by HPLC-DAD-ESI-QTOF-MS. *Food Chemistry* 141:2269–2277.

Andrade PB, Mendes G, Falvo V, Valentão P and Seabra RM (2001) Preliminary study of flavonols in port wine grape varieties. *Food Chemistry* 73:397–399.

Azuma A, Yakushiji H, Koshita Y and Kobayashi S (2012) Flavonoid biosynthesis-related genes in grape skin are differentially regulated by temperature and light conditions. *Planta* 236:1067–1080.

Calderón-Montaño JM, Burgos-Morón E, Perez-Guerrero C and López-Lázaro M (2011) A review on the dietary flavonoid kaempferol. *Mini-Reviews in Medicinal Chemistry* 11:298–344.

Castillo-Muñoz N, Gómez-Alonso S, García-Romero E, Gómez MV, Velders AH and Hermosín-Gutiérrez I (2009) Flavonol 3-O-glycosides series of *Vitis vinifera* Cv. Petit Verdot red wine grapes. *Journal of Agricultural and Food Chemistry* 57:209–219.

Castillo-Muñoz N, Gómez-Alonso S, García-Romero E and Hermosín-Gutiérrez I (2010) Flavonol profiles of *Vitis vinifera* white grape cultivars. *Journal of Food Composition and Analysis* 23:699–705.

Castillo-Muñoz N, Gómez-Alonso S, García-Romero E and Hermosn-Gutiérrez I (2007) Flavonol profiles of *Vitis vinifera* red grapes and their single-cultivar wines. *Journal of Agricultural and Food Chemistry* 55:992–1002.

Cheynier V and Rigaud J (1986) HPLC separation and characterization of flavonols in the skins of *Vitis vinifera* var. Cinsault. *American Journal of Enology and Viticulture* 37:248–252.

Cuyckens F and Claeys M (2004) Mass spectrometry in the structural analysis of flavonoids. *Journal of Mass Spectrometry* 39:1–15.

Cuyckens F and Claeys M (2005) Determination of the glycosylation site in flavonoid mono-O-glycosides by collision-induced dissociation of electrospray-generated deprotonated and sodiated molecules. *Journal of Mass Spectrometry* 40:364–372.

De Quirós AR-B, Lage-Yusty MA and López-Hernández J (2009) HPLC-analysis of polyphenolic compounds in Spanish white wines and determination of their antioxidant activity by radical scavenging assay. *Food Research International* 42:1018–1022.

De Rosso M, Panighel A, Dalla Vedova A, Gardiman M and Flamini R (2015) Characterization of non-anthocyanic flavonoids in some hybrid red grape extracts potentially interesting for industrial uses. *Molecules* 20:18095–18106.

De Rosso M, Tonidandel L, Larcher R, Nicolini G, Dalla Vedova A, De Marchi F, Gardiman M, Giust M and Flamini R (2014) Identification of new flavonols in hybrid grapes by combined liquid chromatography-mass spectrometry approaches. *Food Chemistry* 1635:244–251.

Di Stefano R and Flamini R (2008) High performance liquid chromatography analysis of grape and wine polyphenols. In: *Hyphenated Techniques in Grape & Wine Chemistry*, John Wiley & Sons Ltd, Hoboken, NJ, USA, pp. 33–80.

Downey MO, Dokoozlian NK and Krstic MP (2006) Cultural practice and environmental impacts on the flavonoid composition of grapes and wine: A review of recent research. *American Journal of Enology and Viticulture* 57:257-268.

Downey MO, Harvey JS and Robinson SP (2003) Synthesis of flavonols and expression of flavonol synthase genes in the developing grape berries of Shiraz and Chardonnay (*Vitis vinifera* L.). *Australian Journal of Grape and Wine Ressearch* 9:110–121.

Fang F, Li J-M, Pan Q-H and Huang W-D (2007) Determination of red wine flavonoids by HPLC and effect of aging. *Food Chemistry* 101:428–433.

Figueiredo-González M, Martínez-Carballo E, Cancho-Grande B, Santiago JL, Martínez MC and Simal-Gándara J (2012) Pattern recognition of three *Vitis vinifera* L. red grapes varieties based on anthocyanin and flavonol profiles, with correlations between their biosynthesis pathways. *Food Chemistry* 130:9–19.

Flamini R and De Rosso M (2008) Polyphenols analysis by liquid-mass spectrometry. In: *Hyphenated Techniques in Grape & Wine Chemistry*, John Wiley & Sons Ltd, Hoboken, NJ, USA, pp. 81–128.

Flamini R, De Rosso M and Bavaresco L (2015) Study of grape polyphenols by liquid chromatography-high-resolution mass spectrometry (UHPLC/QTOF) and suspect screening analysis. *Journal of Analytical Methods in Chemistry* Article ID 350259, doi:10.1155/2015/350259.

Flamini R, Mattivi F, De Rosso M, Arapitsas P and Bavaresco L (2013) Advanced knowledge of three important classes of grape phenolics: anthocyanins, stilbenes and flavonols. *International Journal of Molecular Sciences* 14:19651–19669.

Fuhrman B and Aviram M (2002) Polyphenols and flavonoids protect LDL against atherogenic modifications. In: *Handbook of Antioxidants*, Marcel Dekker, Inc, New York, USA.

Ghiselli A, Nardini M, Baldi A and Scaccini C. (1998) Antioxidant activity of different phenolic fractions separated from an Italian red wine. *Journal of Agricultural and Food Chemistry* 46:361–367.

Gichi X (2011) Gene mutations metabolites in response to pathogen. *Bulletin of Applied and Research Science*. Online ISSN 1800-556X, Journal's URL: http://www. barsjournal.net, 1:6–10.

Gómez-Alonso S, García-Romero E and Hermosín-Gutiérrez I (2007) HPLC analysis of diverse grape and wine phenolics using direct injection and multidetection by DAD and fluorescence. *Journal of Food Composition and Analysis* 20:618–626.

Gutiérrez IH, Lorenzo ESP and Espinosa AV (2005) Phenolic composition and magnitude of copigmentation in young and shortly aged red wines made from the cultivars, Cabernet Sauvignon, Cencibel, and Syrah. *Food Chemistry* 92:269–283.

Hermosín-Gutiérrez I, Castillo-Muñoz N, Gómez-Alonso S and García-Romero E (2012) Flavonol profiles for grape and wine authentication. In: *Progress in Authentication of Food and Wine*. Ebeler S. E., Takeoka G. R., Winterhalter P., Eds.; American Chemical Society, Washington, DC, USA, pp. 113–129.

Hernández T, Estrella I, Carlavilla D, Martín-Álvarez PJ and Moreno-Arribas MV (2006) Phenolic compounds in red wine subjected to industrial malolactic fermentation and ageing on lees. *Analalytica Chimica Acta* 563:116–125.

Hilbert G, Temsamani H, Bordenave L, Pedrot E, Chaher N, Cluzet S, Delaunay J-C, Ollat N, Delrot S, Mérillon J-M, Gomès E and Richard T (2015) Flavonol profiles in berries of wild *Vitis* accessions using liquid chromatography coupled to mass spectrometry and nuclear magnetic resonance spectrometry. *Food Chemistry* 169:49–58.

Hvattum E and Ekeberg D (2003) Study of the collision-induced radical cleavage of flavonoid glycosides using negative electrospray ionization tandem quadrupole mass spectrometry. *Journal of Mass Spectrometry* 38:43–49.

Koyama K, Ikeda H, Poudel PR and Goto-Yamamoto N (2012) Light quality affects flavonoid biosynthesis in young berries of Cabernet Sauvignon grape. *Phytochemistry* 78:54–64.

Liang N-N, He F, Bi H-Q, Duan C-Q, Reeves MJ and Wang J (2012). Evolution of flavonols in berry skins of different grape cultivars during ripening and a comparison of two vintages. *European Food Research and Technology* 235:1187–1197.

Liang Z, Yang Y, Cheng L and Zhong G-Y (2012) Polyphenolic composition and content in the ripe berries of wild *Vitis* species. *Food Chemistry* 132:730–738.

Macheix JJ and Fleuriet A (1990) *Fruit Phenolics*. CRC Press, Inc. Boca Raton, Florida, USA.

Makris DP, Kallithraka S and Kefalas P (2006) Flavonols in grapes, grape products and wines: Burden, profile and influential parameters. *Journal of Food Composition and Analysis* 19:396–404.

Masa A, Vilanova M and Pomar F (2007) Varietal differences among the flavonoid profile of white grape cultivars studied by high performance liquid chromatography. *Journal of Chromatography A* 1164:291–297.

Mattivi F, Guzzon R, Vrhovsek U, Stefanini M and Velasco R (2006) Metabolite profiling of grape: Flavonols and anthocyanins. *Journal of Agricultural and Food Chemistry* 54:7692–7702.

Montealegre RR, Peces RR, Vozmediano JC, Gascueña JM and Romero EG (2006) Phenolic compounds in skins and seeds of ten grape *Vitis vinifera* varieties grown in a warm climate. *Journal of Food Composition and Analysis* 19:687–693.

Owens DK, Alerding AB, Crosby KC, Bandara AB, Westwood JH and Winkel BSJ (2008) Functional analysis of a predicted flavonol synthase gene family in Arabidopsis. *Plant Physiology* 147:1046–1061.

Padilla E, Ruiz E, Redondo S, Gordillo-Moscoso A, Slowing K and Tejerina T (2005) Relationship between vasodilation capacity and phenolic content of Spanish wines. *European Journal of Pharmacology* 517:84–91.

Panighel A, De Rosso M, Dalla Vedova A and Flamini R (2015) Putative identification of new *p*-coumaroyl glycoside flavonoids in grape by ultra-high performance liquid chromatography/high-resolution mass spectrometry. *Rapid Communications in Mass Spectrometry* 29:357–366.

Rodríguez-Bernaldo de Quirós A, Lage-Yusty MA and López-Hernández J (2009) HPLC-analysis of polyphenolic compounds in Spanish white wines and determination of their antioxidant activity by radical scavenging assay. *Food Research International* 42:1018–1022.

Souquet J-M, Labarbe B, Le Guernevé C, Cheynier V and Moutounet M (2000) Phenolic composition of grape stems. *Journal of Agricultural and Food Chemistry* 48:1076–1080.

Stecher G, Huck CW, Popp M and Bonn GK (2001) Determination of flavonoids and stilbenes in red wine and related biological products by HPLC and HPLC-ESI-MS-MS. *Fresenius Journal of Analytical Chemistry* 371:73–80.

Tomás-Barberán FA and Ferreres F (2012) Analytical methods of flavonols and flavones. In: *Analysis of Antioxidant-Rich Phytochemicals*, Xu Z., Howard L. R., Eds.; John Wiley & Sons Ltd, Hoboken, NJ, USA.

Trousdale E and Singleton VL (1983) Astilbin and engeletin in grapes and wines. *Phytochemistry* 22:619–620.

Turnbull JJ, Nakajima J, Welford RWD, Yamazaki M, Saito K and Schofield CJ (2004) Mechanistic studies on three 2-oxoglutarate-dependent oxygenases of flavonoid biosynthesis: anthocyanidin synthase, flavonol synthase, and flavanone 3β-hydroxylase. *Journal of Biological Chemistry* 279:1206–1216.

Vallverdú-Queralt A, Jáuregui O, Di Lecce G, Andrés-Lacueva C and Lamuela-Raventós RM (2011) Screening of the polyphenol content of tomato-based products through accurate-mass spectrometry (HPLC–ESI-QTOF). *Food Chemistry* 129:877–883.

Vrhovsek U, Masuero D, Gasperotti M, Franceschi P, Caputi L, Viola R and Mattivi F (2012) A versatile targeted metabolomics method for the rapid quantification of multiple classes of phenolics in fruits and beverages. *Journal of Agricultural and Food Chemistry* 60:8831-8840.

Zhang Y, Shi P, Qu H and Cheng Y (2007) Characterization of phenolic compounds in *Erigeron breviscapus* by liquid chromatography coupled to electrospray ionization mass spectrometry. *Rapid Communications in Mass Spectrometry* 21:2971–2984.

Zhu L, Zhang Y and Lu J (2012) Phenolic contents and compositions in skins of red wine grape cultivars among various genetic backgrounds and originations. *International Journal of Molecular Sciences* 13:3492–3510.

In: Kaempferol ISBN: 978-1-63485-828-1
Editors: T.Garde-Cerdán and A.G. Diago ©2016 Nova Science Publishers, Inc.

Chapter 2

FLAVONOLS: ENHANCEMENT BY USING ELICITORS

Rocío Gil-Muñoz*
Murciano Research Institute and Agricultural Development, Murcia, Spain

ABSTRACT

Elicitation is a strategy used to protect plants against pests and diseases and also to enhance the secondary metabolite concentration in plants and cell cultures. Phenolic compounds can be found among these secondary metabolites; being a very spread group with comprise flavonol compounds, and among the flavonols, kaempferol is a very important compound due to its application in medicine, and consequently in human health.

Different elicitors can be used to obtain an enhancement of flavonols, and consequently of kaempferol in different crop plants. Many researchers have reported different experiences about the use of different abiotic and biotic elicitors in several cultivars obtaining contradictory results according to the compound used, the doses and the application moment.

It is said that it is not clear the influence of different elicitors in flavonols composition, and more specifically in kaempferol. Thus, in this chapter a general vision about the use of elicitors and the influence of

* E-mail: mariar.gil2@carm.es.

them in the concentration of phenolic compounds and specifically in flavonols is shown.

Abbreviations

UV (ultra-violet);
MeJ (methyl jasmonate);
BTH (benzothiadiazole);
SAR (systemic acquired resistance);
ROS (reactive oxygen species);
PSM (plant secondary metabolite);
MW (molecular weight);
PAL (phenylalanine ammonia lyase);
ABA (abscisic acid);
CHT (chitosan);
JA (jasmonic acid);
ASM (Acibenzolar-S-methyl);
BABA (β-amino butyric acid);
1-MCP (1 methyl ciclopropene);
ACE (angiotensin I-converting enzyme).

1. Introduction

Plant secondary metabolites (PSM) can be defined as compounds with an unrecognized role in the maintenance of fundamental life processes in the plants but with an important role in the interaction of the plant with its environment. The production of these compounds is often low (less than 1% dry weight) and depends greatly on the physiological and developmental stage of the plant (Dixon, 2001; Oksman-Caldenteny and Inze, 2004).

Among phenolic compounds are a kind of secondary metabolites named flavonols, which present important properties in plants and human health. These compounds can be found in plants in minor quantities. One of these compounds is called kaempferol (3,5,7-trihydroxy-2-(4-hydroxyphenyl)-4H-1-benzopyran-4-one), a yellow compound with a low molecular weight (MW: 286.2 g/mol) that is commonly found in plant-derived foods and in plants used in traditional medicine. Numerous edible plants contain kaempferol, and it has been estimated that the human dietary intake of this polyphenol may be up to approximately 10 mg/day (Gates et al., 2007; Adebanovo et al., 2005; Lin et al., 2007). Epidemiological studies have found a positive association between the consumption of kaempferol containing foods and a reduced risk of developing cardiovascular diseases and some types of cancer. Numerous *in vitro* and some animal studies support a role of kaempferol in the prevention and/or treatment of these and other diseases, such as neurodegenerative diseases, infectious diseases, diabetes, osteoporosis, anxiety, allergies.

To obtain a higher concentration of secondary metabolites in plants, and specifically in flavonols, it is necessary to develop alternatives to the intact plant for the production and enhancement of plant secondary metabolites. The aim can be reached by three main ways: classical breeding methods, genetic manipulation and modification of metabolism by means of elicitors. Many biotechnological strategies have been hypothesized and experimented for enhancing the production of PSM from plants, such as pruning, cluster thinning, or the use of deficit irrigation. Classical breeding methods have been practiced for many years to improve certain characteristics in plants. Plant breeding can be accomplished by different techniques like simply selecting plants with desirable characteristics for vegetative propagation, the deliberate crossing of closely or distantly related individuals to produce new crop varieties or lines (called hybrids) with desirable properties, or selecting clones from the same variety that express interesting attributes. Genetic engineering has been used to modify flavonoid biosynthesis in plant tissues by using different structural flavonoid genes from different plant sources (Martens et al., 2003). In conclusion, the improvement of nutritional and nutraceutical value of plant/sprouts will be beneficial for human health but genetic engineering of plant food has proven to be controversial and therefore modification of chemical composition and plant food selected bioactivities by elicitors is cheaper and socially more acceptable.

The third of the strategies proposed has been called elicitation which is a process used to mimic the natural reactions of plants towards various environmental stresses and has been successfully applied to plants to induce secondary metabolite production (Dixon, 2001), although the effects of the elicitation are highly dependent on the concentration of the elicitor used. Elicitors can also be used by means of the biotechnological production of valuable secondary metabolites in plant cell or organ cultures as an attractive alternative to the extraction of whole plant material. Elicitor which may be defined as a substance which, when introduced in small concentrations to a living cell system, initiates or improves the biosynthesis of specific compounds. Several parameters such as elicitor concentration and selectivity, duration of elicitor exposure, age of culture, cell line, growth regulation, nutrient composition, quality of cell wall materials, substrate enhancement of product accumulation etc, have been reported as relevant parameters in the results obtained (Namdeo, 2007). Screening, selection, elicitation and media optimization are the methods applied for improving production of secondary metabolites in cell cultures. The enhanced production of the secondary metabolites from plant cell cultures through elicitation has opened up a new area of research which could have important economic benefits for pharmaceutical industry. However, the use of plant cell or organ cultures has had only limited commercial success. Most attempts to produce secondary metabolites *in vitro* have failed because the cells have not produced the compound in sufficient quantity and yields have been unpredictably variable. These two factors have usually been considered to be associated with the use of undifferentiated cells (Verpoote, 1998).

Elicitors were first used to increase plant resistance to pathogens although it was found that the mechanism involved increased polyphenol levels. Although until now, fungicide application is the most effective method for controlling plant diseases caused by different pests, legislation is limiting and reducing their use. That's why there are currently studies for the identification of additional and environmentally friendly approaches in the control of their associated diseases. Among these options, it is the use of elicitors whose action method consist in the resistance process, mediated by the accumulation of endogenous salicylic acid (SA), a metabolite downstream the biosynthetic pathway initiated by phenylalanine ammonialyase (PAL). It is called systemic acquired resistance (SAR) and is based on the induction of secondary metabolic pathways and the increased synthesis of products, phenolic compounds among them, as a response to pathogen attack (Iriti et al., 2005). Systemic acquired resistance (SAR) offers the prospect of long-lasting, and broad-spectrum disease control through activation of the resistance defense machinery of the plant itself. Plant activators are products employed in crop protection able to elicit SAR. Therefore, they may trigger the plant own defense in response against pathogen attacks, mainly stimulating mechanisms such as the biosynthesis of phytoalexins, plant secondary metabolites with a broad spectrum biological activity.

However, there are some problems that slow down the exploitation of SAR for crop protection making difficult its implementation in: i) open fields, the effectiveness of some inducers may be variable; ii) consumers, many farmers and crop protectors ask for agricultural products with very high performances; and also due to iii) a number of inducers are not registered as plant protection products, but as biostimulants of plant defenses, and their compositions are not known (e.g., Kendal®, Pom-PK®). In addition, the efficacy of SAR in the field is variable, and it can be influenced by the environment, crop genotype, nutritional status, and the extent to which plants have already been induced (Reglinski et al., 2007; Walters and Fountaine, 2009). Unfortunately, our understanding of the impact of these factors on the expression of SAR is poorly developed.

Despite has just been mentioned above, many researchers have reported about the influence of certain elicitors in the obtaining of higher concentration of flavonoids in plants. Schijlen et al., (2006) were able to produce transgenic tomatoes that accumulated high levels of new phytochemicals such as stilbenes, deoxychalcones, flavones, and flavonols. For instance, in strawberries cultivated in greenhouses, pre-harvest treatment with benzothiadiazole (BTH), has proved to be useful for preventing powdery mildew and for increasing the content of quercetin and kaempferol (Antonnen et al., 2003). Furthermore, the treatment has enhanced the accumulation of ellagic acid, ellagitannins, p-coumaric acid, gallic acid, and kaempferol hexose in leaves, and kaempferol malonylglucoside in fruits (Hukkanen et al., 2007), and also the amount of quercetin and kaempferol in berries (Karjailanen et al., 2002).

The activation of chalcone synthase (CHS), stilbene synthase (STS), glycogen synthase (UDP) glucose: flavonoid-*O*-transferase (UPGT), proteinase inhibitors and chitinase gene expression has also been reported in pre-harvest treatments of grapevine with methyl jasmonate (MeJ). Such activations triggered the accumulation of both stilbenes and anthocyanins in cells (Belhadj et al., 2008). Many studies have confirmed the activation of PAL following postharvest application of the elicitor in lychees (Yang et al., 2011), peaches (Jing et al., 2009), apples, plums, table grapes, strawberries (Heredia et al., 2009) with a subsequent increase of total phenols. In a different fruit, red raspberry, the enhancement in the levels of myricetin, quercetin, and kaempferol has also been reported in postharvest treatment with MeJ (Moreno et al., 2010). A few papers have reported that grape berries also respond to postharvest treatments with high concentrations of ethylene. Bellincontro et al., (2006) showed that *cv. Aleatico* berries react with ethylene in postharvest (500 mg L^{-1} for 15 h) and 1-MCP treatments; both polyphenol and anthocyanin contents and the aromatic quality of the grapes were affected by the treatments. Working on the same aromatic red-skinned variety and with the aim of exploiting ethylene as a postharvest dehydration enhancer, Botondi et al. (2011) showed that, when this gas (at 1000 ppm for 48 h) is applied, the activity of specific cell wall enzymes increases and the resulting anthocyanin concentration is higher in ethylene-treated berries. This treatment was effective in increasing the concentrations of phenols and anthocyanins in wine with some pronounced effects also in volatile compound profile.

In short, the interest in the development of new alternatives, using different elicitors in plants instead of using fungicides in crops, has two important consequences: on the one hand, these allow the use of friendly environmental tools against pest and diseases, and on the other hand, these elicitors lead to an increase in the phenolic compounds of plants, also flavonols compounds, and in consequence to get functional foods by using different strategies which enhance the concentrations of health promoting compounds.

2. Phenolic Compounds as Plants Defense

Polyphenolic compounds are important for both plants and humans for several reasons: their physiological role in plants (they protect plants from biotic and abiotic stress factors), their technological significance for food processing (most of these metabolites are responsible for the organoleptic and qualitative properties of foods originating from such plants) and their nutritional characteristics (most of these metabolites are responsible for the organoleptic and qualitative properties of foods originating from such plants).

2.1. Physiological Role in Plants

Plants cannot run away when attacked by herbivores nor do they have an immune system against bacteria, fungi, or viruses infection. Similar to the situation in other sessile organisms (such as marine animals), plants have developed biologically active secondary metabolites during evolution that help them to defend themselves against herbivores (insects, mollusks, vertebrates), microbes, viruses, and other competing plants (Figure 1) (Swain, 1977; Wink, 1988, 1999; Harborne, 1993; Roberts and Wink, 1998).

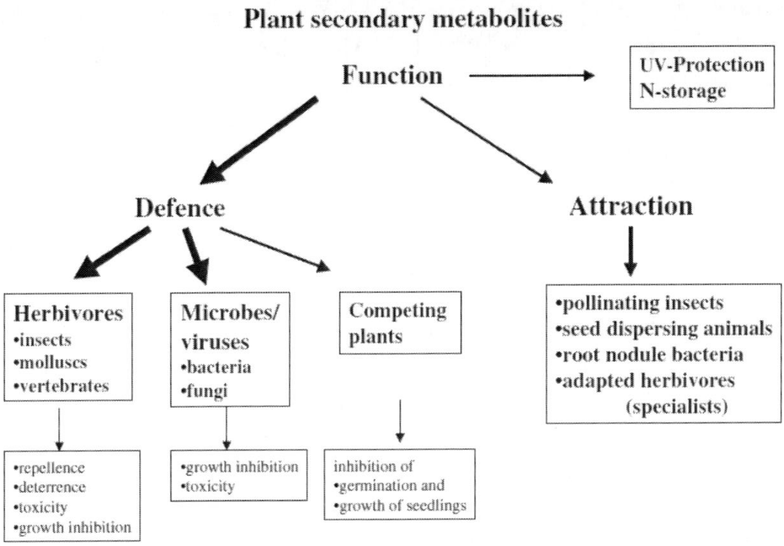

Figure 1. Function of secondary metabolites in plants (Wink, 2006).

Therefore, secondary metabolites act as defense in plants against hervibores, microbes, viruses or competing plants, as well as signal compounds for attracting pollinating or seed dispersing animals and lasting, protecting the plant from ultraviolet radiation and oxidants (Swain, 1997; Kutchan 2001). Since defense against pathogens is one of the functions of flavonoids in plants, also for this reason plants with improved flavonoid production have been made (Yu et al., 2003; Fischer et al., 2007; Jeandet et al., 2002). Their concentrations are also affected by several other abiotic factors like temperature, drought, salinity, seasonality, circadian rhythm, altitude, light, UV radiation, metal ions, wounds and nutrient deficiencies, among others (Gouvea et al., 2012).

Phenolic compounds play many physiological roles in plants including reproduction, growth and defense against biotic and abiotic stresses (Achakzai et al., 2009) indeed, some of these phenolic compounds are only induced when stress factors are present, among them, the so-called phytoalexins, which are specifically involved in defense mechanisms and are synthesized after pathogen or predator attacks (Ruiz-García and Gómez-Plaza, 2013). They also have allelopathic activity by which they reduce the growth of neighboring plants (Taiz and Zeiger, 2006). It has been found that phenols can also be used as a stress indicator because they are increased by exposure of toxic chemicals and stresses in the plants (Siddiqui and Zaman, 2004; Achakzai et al., 2009).

Among the major phenolic compounds are the flavonoids (anthocyanins, flavones, flavonols and isoflavonoles) which have a crucial role in plant growth and defense mechanism against microbes and insects (Taiz and Zeiger, 2006). Furthermore, on the other hand, due to their significance in plant defense and in human health, the main purpose is to discover how phenolic compounds can be enhanced in crops. Besides genetic transformation (forbidden in most countries), a wide range of factors are able to modify the grape phenolic content, including agronomic practices, clonal selection, and those stress factors that may trigger SAR establishment (Parr and Bolwell, 2000). However, it has been demonstrated that SAR can also be induced or enhanced by the exogenous application of natural or synthetic compounds that may have powerful effects (Scarponi et al., 2001).

They may trigger the plant own defense response against attacks, such as the biosynthesis of plant secondary metabolites (Bassi, 2011). The induction of biosynthesis of phenolic compounds and their higher concentration is also dependent upon the response to pathogen attack (biotic factors) to enable the host plants against defense.

Different levels of phenolic compounds appear in response to environmental factors like nutrient availability and light intensity, etc (Naghiloo et al., 2012). The concentration and accumulation of phenolic compounds are also affected at genetic level. The increase in the production of different phenolic compounds can be achieved through the different strategies. Genetically, it has been noticed that secondary metabolites does not remain stabilized and they can be influenced by several factors. The factors that influences in fluctuation of PSMs are the following: a) genetic, b) ontogenetic, c) morphogenetic and d) environmental factors as it can be seen in Figure 2.

Also, the increase in production of different phenolic compounds can be achieved through *in vitro* tissue culture by using elicitors and plant growth regulators which clearly demonstrate that their concentration can be altered.

2.2. Technological Significance for Food Processing

Most of the phenolic compounds are responsible for the organoleptic properties of foods. For example, anthocyanins, constitute a pigment group responsible for the color of a great variety of fruits, flowers and leaves (Harborne and Williams, 2000), and flavan-3-ols are polyphenols involved in the bitterness and astringency of tea, grapes and wine (Noble, 1994; Halsam and Lilley, 1998; Gonzalo-Diago et al., 2014).

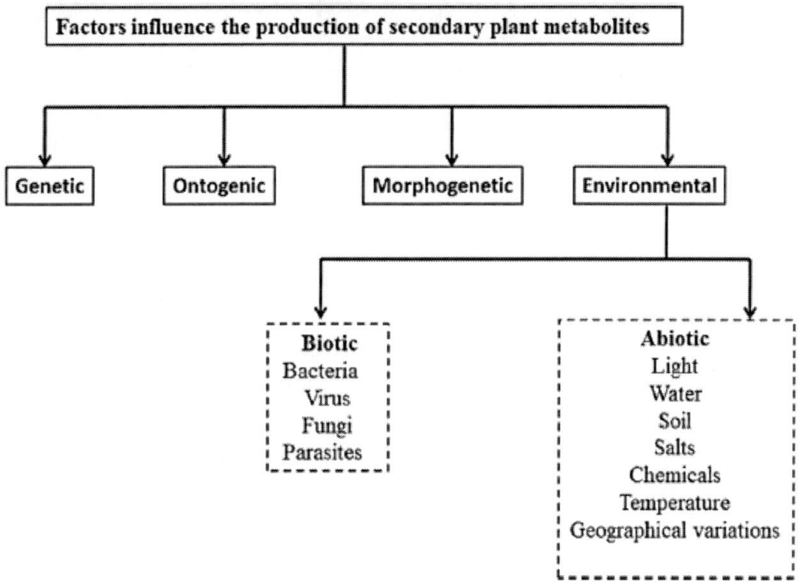

Figure 2. Factors affecting the synthesis of plant secondary metabolites (Verna and Shukla, 2015).

Phenolic compounds contribute significantly to imparting specific flavours and colours to various plants widely used in foods and beverages. Other examples includes capsaicin, responsible for the pungent properties of the red peppers, alkylphenols, responsible for the characteristic taste and odour of clove oil, tannins, which add a distinct bitterness or astringency to the taste of certain foods, and the anthocyanin pigments, such as the pelargonidins, the cyanidins and the delphinidins (responsible for red, blue and purple colours) (Croteau et al., 2000). The glucosinolates, characteristics of cruciferous foods, also add bitter taste (progoitrin) and aroma intensity (total glucosinolates) to vegetables (Passini et al., 2011). Finally, the plants fitness greatly benefits from secondary metabolites with confer colors and scents to flowers and fruits, thus playing important roles in reproduction (Harborne, 2001).

2.3. Nutritional Characteristics

The phenolic compounds are unique sources for the industry in the form of food additives and pharmaceuticals products. Since a long time ago, it is known that the composition of secondary metabolites has a great effect in the quality and health potential of food and food products (Stobiecki et al., 2002). For this reason, there is currently a growing interest in the development of strategies to enhance the level and composition of flavonoids in plants.

Many studies have suggested that a high intake of polyphenol-rich foods may have cardiovascular benefits, and provide some level of cancer chemo preventive activities and beneficial effects against other less prevalent but devastating illnesses, such as Alzheimer's disease and urinary bladder dysfunctions (Leifert and Abeywardena, 2008; Croizier et al., 2009; De Pascual-Teresa et al., 2010; Pezzuto, 2008). Phenols act as antioxidants through different mechanisms (Parr and Bolwell, 2000): (1) hydroxyl groups with electrons of the phenyl can capture free radicals; (2) the generation of free radicals catalyzed by metals is diminished since they chelate metallic ions; (3) the cycle of generating new radicals is stopped through the donation of a proton from the phenolic compounds to the radicals and (4) polyphenols inhibit pro-oxidant enzymes that generate free radicals, such as lipoxygenases, cyclo-oxygenases and xanthine oxidase. The antioxidant activity of food phenolic compounds is of nutritional interest, since it has been associated with the potentiation of the promoting effects of human health through the prevention of several diseases; thus play a considerable role in reducing the risk of cancer humans (Sharma et al., 2012). Additionally, in some cases, these compounds may also be used with therapeutic purposes due to their pharmacological properties. However, the antioxidant activity of phenolic compounds depends largely on the chemical structure of these substances (Pérez-Lamela et al., 2007). Among the phenolic compounds with known antioxidant activity, flavonoids, tannins chalcones and coumarins as well as phenolic acids are highlighted.

3. ELICITORS

In agriculture, plant varieties were domesticated and over time bred for yield and fruit quality. As a consequence, plant disease resistance is often decreased compared to wild varieties. Most crops are susceptible to numerous diseases caused by different microorganisms (pathogens). Plant diseases were responsible for severe economic and nutritional crises and still currently responsible for a considerable loss in the worldwide crop production. To date, ensuring a satisfactory yield and the quality of the harvest requires an extensive use of numerous phytochemical pesticides. However, pesticides harm crops, the environment, even the health of farmers and consumers. For these reasons alternative and sustainable disease management is required. Alternatives include organic and integrated farming practices, biological control, the use of resistant hybrids or transgenic crops. However, some national legislative bodies do not allow genetic crop improvement by transgenesis and assisted crossing may also be prohibited for some crops, such as wine, protected by appellation seals. The strategy of induced resistance represents one alternative compatible with organic-farming. It consists of stimulating the plant immune system with elicitors, natural molecules that mimic a pathogen attack or a danger state, or by living organisms. Induced resistance may represent an interesting strategy for crops when fungicide-provided control is undesired (Burketová et al., 2015).

Elicitation consists in a process used to mimic the natural reactions of plants towards various environmental stresses and has been successfully applied to several cultivars and plant cell culture to induce secondary metabolite production (Dixon, 2001). Consequently, elicitation is also a good strategy to induce the synthesis of different classes of bioactive secondary metabolites. The efficacy of this treatment depends mainly on plant genetics, elicitor nature and on the concentration of the elicitor used (Baenas et al., 2014). Several studies have used pre-elicitor treatments to affect the phenolic composition of different sprouts and only few of them have focused on legume seeds (Burguieres et al., 2007; Limón et al., 2014; Liu et al., 2013).

Plants treated with elicitors generally develop resistance to host, because application of elicitors on plant surface activates multiple signaling pathways of intracellular defense (Odjacova and Hadjiivanova, 2001; Garcia-Brugger et al., 2006; Bent and Mackey, 2007; Holopainen et al., 2009). Elicitors are very stable molecules that induce an immune defense response in plants, they have low molecular weight and synthesized as such or released from polymeric precursors during infection (Ozeretskovskaya and Vasyukova, 2002; Zhao et al., 2005; Boller and Felix. 2009; Holopainen et al., 2009). In a broad sense, “elicitors,” for a plant refers to chemicals from various sources that can trigger physiological and morphological responses and phytoalexin accumulation. Elicitor needs to be recognized by a plant receptor (protein), which activates the expression of defense genes. There are highly diverse molecules both in nature and origins.

Resistance responses are triggered following the recognition of a range of biological, chemical and physical factors named as “elicitors” (Terry and Joyce, 2004). Studies about exogenous application of salicylic acid (SA) in many fruit crop, such as sweet cherry (Yao and Tian, 2005; Bi et al., 2007), mango (Zeng et al., 2006) and pear (Tian et al., 2006), have allowed to identify SA as a natural inducer of plant defenses. Acibenzolar-S-methyl ester (ASM, BTH, actigard®, Bion®) a functional analogue of SA, activates disesase resitance in crops to a wide variety of pathogens (Hammerschmidt et al., 2001). The resistance induced in strawberry (Terry and Joyce, 2004), peach (Liu et al., 2005), Yali pear (Cao et al., 2005) and pear (Tsiantos et al., 2003) has protected fruits from postharvest pathogens.

Also, fruit losses due to postharvest fungal rots could be reduced if the infections could be kept in their quiescent phase for extended periods during storage and marketing. One possible way to prolong quiescence is to maintain the natural antifungal barrier present in the unripe fruit at an inhibitory level into the post-climacteric phase. Since long time ago, induction of natural resistance has been considerable interesting in the control of postharvest disease (Adikaram, 1990; Yang et al., 2007).

The production of PSMs depends not only on genetic regulation but also on environmental factors. These factors can be used as elicitor treatments to increase the production of plant bioactive compounds (Jansen et al., 2008; Ncube et al., 2012).

As it has been explained in the previous section, the interest in developing functional foods by using elicitors with enhance the concentrations of health promoting compounds has been increased in recent years. Several strategies have been implemented to enhance the level of phenolic compounds in the crop plants. One method tested tries to support better agricultural practices to produce genetically modified plants (Ellis et al., 1994). However, the strategies which can help in reducing agrochemicals in crops commercial interest and provide a more economical and environment friendly agriculture are currently in demand. Recently, the use of elicitors has shown to be an effective way to enhance growth and secondary metabolite production in crop plants (Singh, 2016). In Table 1 can be observed different examples for the control of fungal plant diseases by means of elicitors.

Physical elicitors include, for example, high and low temperatures, and ultraviolet and gamma radiation. Chemical elicitors, such as chitosan, benzothiadiazole (BTH), harpin, and 1-methylciclopropane, among others, are agrochemicals that can mimic the action of the signaling molecules SA and jasmonic acid (JA) and their derivates, or simulate the attack of a pathogen. These molecules may interact with receptors in the plant, activating defense responses and triggering, in some cases, a hypersensitive reaction. Enzymes of the phenylpropanoid biosynthetic pathway (phenylalanine ammonia lyase and chalcone isomerase) were observed to accumulate after the application of exogenous BTH (Gozzo, 2003) and MeJ (Repka, 2001), since the induction of secondary metabolite accumulation is an important stress response and jasmonates and SA (or its analogues) function as necessary signaling molecules (Tassoni et al., 2005).

There are two groups of elicitors on the basis of their origin, the biogenics and the abiogenics. The first ones are divided into two groups, exogenous and endogenous. Exogenous elicitors are substances originated outside the cell like polysaccharides, polyamines and fatty acids whereas endogenous elicitors are substances originated inside the cell like galacturonide or hepta-glucosides, etc. The exogenous are isolated from pathogens or culture medium, whereas endogenous tissues are isolated from same plant (Bent and Mackey, 2007).

Also, elicitors can be classified into two groups based on their nature, namely abiotic and biotic elicitors. Abiotic elicitors are substances of non-biological origin, predominantly inorganic salts, and physical factors, such as UV radiations, alkalinity, temperature, osmotic pressure and heavy metal ions, whereas biotic elicitors are substances with biological origin; they include polysaccharides derived from plant cell walls (pectin or cellulose) and micro-organisms (chitin or glucans) and glycoproteins or G-protein or intracellular proteins whose functions are coupled to receptors and act by activating or inactivating a number of enzymes or ion channels (Veersham, 2004). Table 2 shows the classification of elicitors.

In the next section, the most studied elicitors in different crops plants (fruits and vegetables) will be described: methyl jasmonate (MeJ), benzothiadiazole (BTH), chitosan (CHT), UV-radiation, cell wall yeasts, harpine, and abscisic acid.

3.1. Methyl Jasmonate (MeJ)

MeJ is a phytohormone, a naturally occurring plant growth regulator, that plays an important role in effectively suppress some important diseases in fruits (Santos-Buelga and Scalbert, 2000; Tzortzakis, 2007). MeJ is a compound derived from jasmonic acid (JA) with a molecular weight of 224.3. It has similar activity to JA in plants and thus it is able to activate the enzymes responsible for the biosynthesis of polyphenols, such as PAL enzyme. Exhibiting the characteristics of MeJ, the application of exogenous MeJ increases the content of secondary metabolites in various plants. Also, it has been reported that it increases the content of a number of bioactive plant components.

Table 1. Examples of induced resistance for the control of fungal plant diseases in the field (Bassi, 2011)

Disease	Causal agent	Crop	Inducer	References
Blue mold	*Penicillium expansum*	Apple	Quercetin, umbelliferone	Sanzani *et al.*, 2008
			trans-2-hexenal	Neri *et al.*, 2006
			BTH, BABA, methyl jasmonate	Quaglia *et al.*, 2009
		Grapes	Chitosan	Romanazzi *et al.*, 2006, 2009
		Pear	*trans*-2-hexenal	Neri *et al.*, 2006
		Peach	BTH	Liu *et al.*, 2005
Brown rot	*Monilinia laxa*, *M. fructigena*, *M. fructicola*	Sweet cherry	Chitosan	Romanazzi *et al.*, 2003
Gray mold	*Botrytis cinerea*	Apple	BTH, BABA, methyl jasmonate	Quaglia *et al.*, 2009
		Grapes	Chitosan	Romanazzi *et al.*, 2002, 2006, 2009
		Strawberry	Chitosan	El Ghaouth *et al.*, 1992; Zhang and Quantick, 1998; Reddy *et al.*, 2000; Romanazzi *et al.*, 2000
			Chitosan, BTH, Fitocalcio®, Algition®	Santini *et al.*, 2009
		Sweet cherry	Chitosan	Romanazzi *et al.*, 2003
		Grapes	BTH	Iriti *et al.*, 2004, 2005
		Tomato	BTH	Iriti *et al.*, 2007
Green mold	*Penicillium digitatum*	Grapefruit	BABA	Porat *et al.*, 2003
Rhizopus rot	*Rhizopus stolonifer*	Strawberry	Chitosan	Romanazzi *et al.*, 2000
			Chitosan, BTH, Fitocalcio®, Algition®	Santini *et al.*, 2009

Table 2. Classification of elicitors for the synthesis of secondary metabolites (Namdeo, 2007)

A) Nature of Elicitor	
Biotic elicitors	**Abiotic elicitors**
- Directly released by microorganisms and recognized by the plant cell (enzymes, cell wall fragments) - Formed by action of microorganisms on plant cell wall (fragments of pectins etc.) - Formed by the action of plant enzymes on microbial cell walls (chitosan, glucans) - Compounds, endogenous and constitutive in nature, formed or released by the plant cell in response to various stimuli.	- Of physical or chemical nature working via endogenously formed biotic elicitors - UV light - Windfall - Denatured proteins (RNase) - Freezing and thawing cycles - Non essential components of media (agarose, tin, etc.) - Heavy metals - Chemicals - With high affinity to DNA - With membrane-destroying activities like detergents: xenobiochemicals - Fungicides (Maneb, Butylamine, Benomyl) - Herbicides (Acifluorofen)
B) Origin of Elicitor	
Exogenous elicitors	**Endogenous elicitors**
- originated outside the cell, including the reaction immediately or via endogenous mediators - Polysaccharides: Glucomannose, Glucans, Chitosan - Peptides as poly cations: Monilicolin, Poly-L-lysine, Polyamines, Glycoproteins - As enzymes: Polygalacturonase, Endo-polygalacturonic acid lyase, Cellulase - Fatty acids: Arachidonic acid, Eicosapentanoic acid	- formed via secondary reactions induced by a signal of biotic or abiotic nature in the cell - dodeca-β-1,4-D-galacturonide - hepta-β-glucosides - alginate oligomers

There are many studies concerning the impact of exogenous MeJ on flavonoid content, antioxidant capacity and post-harvest life of various berries (Ayala-Zabala et al., 2004; Wang et al., 2009; de la Peña-Moreno et al., 2010b). In recent years, phytochemical contents and other quality parameters have been improved with pre- and post-harvest plant growth regulator treatments (Khan and Singh, 2007; Lara, 2013). MeJ is among the most common plant growth regulators. Since it has a regulatory role in fruit growth and ripening, it can easily affect the ripening processes of both climacteric and no climacteric fruits (Lalel et al., 2003; Pena-Cortes et al., 2005).

Different works have reported the results obtained when MeJ was used alone or in combination with other compounds. In both cases, different results were shown for the different crop plants. For instance, it was reported in previous studies that MeJ had significant impact on fruit peel colour development, anthocyanin accumulation, phenolic compounds and antioxidant activity of the fruits (Rudell et al., 2005, Shafiq et al., 2013). Wang and Zheng (2005) reported an increase in flavonoids, total phenolics and antioxidant activity of raspberry with MeJ treatments. Similarly, significant increases were reported with pre-harvest MeJ treatments in the phytochemical content of blueberries (Percival and Mackenzie, 2007), blackberries (Wang et al., 2008), apples (Shafiq et al., 2011) and grapes (Ruiz-García et al., 2012). On the other hand, Khan and Sing (2007) reported increased phytochemical contents of Japanese plums (*Black Amber, Amber Jewel* and *Angelino*) with post-harvest MeJ treatments. This elicitor has been also shown to stimulate secondary metabolites such as stilbenes in leaves and berries of grapevines (Larrondo et al., 2003); to enhance anthocyanin accumulation in soybean seeds (Francheschi and Grimes, 1991), peach shoots (Saniewski et al., 1998) and strawberry ripening (Pérez et al., 1997); and to increase synthesis of alkaloids, terpenoids and phenolics. Finally, investigations about the effect of MeJ on berry composition have proven that MeJ affects the antioxidant capacity, quality parameters, aroma compounds and shelf-life of fruits (Ghasemnezhad and Javaherdashti, 2008; Blanch et al., 2011; Wang and Zheng, 2005).

Other studies have shown the effect of MeJ application in combination with other elicitors showing a synergic effect. For example, Heredia and Cisneros-Zevallos (2009) reported that the exposition to MeJ and ethanol promotes the accumulation of phenolic compounds by triggering the phenylpropanoid metabolism through an increase in PAL activity in carrots. De la Peña-Moreno et al., (2010a) showed an enhancement in biological properties of raspberries caused by MeJ in conjunction with ethanol treatment.

This combination provides more antioxidant properties to regular diets, at low cost, and can be an alternative to genetic modifications and breeding activities.

In most cases, methyl jasmonate (MeJ) and salicylic acid (SA) have been shown to be potent elicitors, which help to trigger the signal transduction pathway responsible for biochemical and physiological processes in plants (Creelman and Mullet, 1997). The use of MeJ and SA elicitation method was reported for the production of anthocyanins in soybean (Franceschi and Grimes, 1991), phenolics in sweet basil (Singh et al., 1998), alkaloids in *Nicotiana* species) and terpenoids in *Hyoscyamus muticus* (Martin et al., 2002).

3.2. Benzothiadiazole (BTH)

BTH is a functional analogue of the plant endogenous hormonelike compound salicylic acid, that, in untreated plants is required for the induction of defense genes leading to a broad spectrum, and long lasting systemic immunity (SAR) (Iriti et al., 2005). It has been proven that BTH is effective against downey mildews (Iriti et al., 2003). In the strawberry, BTH has been used to control gray mold and *Phytophthora* (Terry and Joyce, 2004) but not powdery mildew despite the importance of the disease and the efficacy of BTH found against powdery mildews in other plant species (Görlach et al., 1996).

Acibenzolar-S-methyl (ASM), named also benzothiadiazole (BTH) is an efficient broad-spectrum resistance inducer against bacterial, fungal and viral diseases in different monocot and dicot crops (Walters et al., 2013). It is commercialized in different forms known as Bion (in Italy) or Actigard (Syngenta) and it is widely used in agriculture. BTH has been registered in several countries and it is extensively applied to protect a number of crops from fungal diseases (Walters and Fountaine, 2009). ASM/BTH induces responses identical to SA-activated defense and leads to SAR (Ryals et al., 1996). BTH was shown to act as SA analogous since it is able to activate SAR in NahG transgenic plants, which lacks SA due to over expression of SA degrading the enzyme salicylate hydroxylase. The usual active dose used in various studies is 0.03–0.4 mM. ASM/BTH also has a direct fungitoxic effect and reduces yield (Gozzo and Faoro, 2013; Walters et al., 2013). In a screen of various benzothiadiazole derivatives, benzo-1,2,3-thiadiazole-7-carbothioic acid S-methyl ester (BTH, acibenzolar-S-methyl) emerges as a strong inducer of SAR in numerous plant-pathogen combinations, with much lower phytotoxicity than either SA or 2,6-dichloroisonicotinic acid (INA) (Friedrich et al., 1996; Schurter et al., 1987).

Different researchers have shown the results obtained with BTH in different crops. Postharvest treatments in bananas and mangoes also resulted in the activation of polyphenol oxidase (PPO) and peroxidases (POD) and an increased of total phenolic content (Zhu and Ma, 2007; Li et al., 2011). However, the effect on flavonoid metabolism might be species-dependent since PAL was inhibited, whereas POD and PPO were activated by postharvest BTH treatment in loquat (Zhu et al., 2007).

3.3. Chitosan

Chitosan (CHT) is a natural and low cost polymer deacetylated chitin derivate, and it is a widely used elicitor molecule in crop protection. Chitin is a major component of fungal cell walls and is also present in the cuticle of non-vertebrates such crustacean shells, insect exoskeletons, in egg shells of nematodes, protists and algae (Bueter et al., 2013). Naturally occurring chitin is not a pure homopolymer, but is a heteropolymer with a varying degree of deacetylation and a different content of glucosamine. Its effectiveness is higher when molecular weight is between 10 and 100 kD and the deacetylation degree range from 80 to 90 percent. It is mainly composed of glucosamine, 2-amino-2-deoxy-D-glucose (Freepons, 1991) and it is widespread in nature being the second most abundant carbohydrate on earth. CHT is found in many fungal species such as *Cryptococcus* (Baker et al., 2007), *Rhizopus, Absidia and Mucor* (Miyoshi et al., 1992).

The positive charge of CHT confers to this polymer numerous and unique physiological and biological properties with great potential in a wide range of industries such as cosmetic (lotions, facial and body creams) (Lang and Clausen, 1989), food (coating, preservative, antioxidant, antimicrobial) (Benjakul et al., 2000), biothecnology (chelator, emulsifier, flocculent) (Sandford, 1989), pharmacology and medicine (fibers, fabrics, drugs, artificial organs) (Liu et al., 2001) and agriculture (soil modifier, films, fungicide, elicitor) (Ren et al., 2001).

Different examples can be shown about the effect of chitosan over different disease crops. The treatment with chitin reduced the susceptibility of rice to *M. oryzae*. The effect of chitin treatment on resistance of fungal pathogens is different according to the diverse plant species. Low molecular weight oligochitosan induced protection of grapevine leaves against *Botrytis cinerea*. The elicitor activity of CHT was first demonstrated in a screen of fungal cell wall components that were assayed for their ability to induce phytoalexin accumulation in pea pods and induce resistance to the fungal pathogen *Fusarium solani* (Hadwiger, 1980). Also, CHT induced resistance is associated with an increased content of polyphenolic phytoalexins in treated plant tissues, because of the stimulation of phenylpropanoid pathway, the biosynthetic route leading to polyphenol synthesis. Elicitation of this metabolic pathway by CHT has been reported in grape and other plants and correlated with the increase in both activity and transcript levels of phenylalanine ammonia lyase and chalcone synthase, key enzymes of the phenylpropanoid route (Nandeeshkumar et al., 2008). These effects are also related with the improvement of quality and antioxidant power of foodstuffs derived from CHT-treated crops (Iriti et al., 2011; Cho et al., 2008).

Many pre- and post-harvest treatments with chitosan have demonstrated that this compound can activate the enzyme PAL and increase total polyphenols in table grapes, controlling storage gray mold (Romanazzi et al., 2002), and activating PPO (Meng et al., 2010). It may also enhance the activity of defense-related enzymes in bananas (Meng et al., 2012) and increase the amount of total polyphenols in strawberries (Mazaro et al., 2012). In addition, chitosan has proved to be effective at controlling powdery mildew and at increasing the total polyphenol content of grapes. Moreover, wines made from chitosan-treated grapes showed a higher total polyphenol content and antiradical power than those made from fungicide-treated and untreated grapes (Iriti et al., 2011).

3.4. UV-Radiation

UV-B radiation (280-315nm) is an important environmental factor that may induce photobiological stress in plants, affects plant growth and development, activates the plant defense system, and in many cases, induces the production of secondary metabolites (Schreiner et al., 2014). This radiation could either stimulate protective mechanisms or activate repair mechanisms to cope with the UV-B stress, the most common protective mechanisms against potential damaging irradiation is the biosynthesis of secondary metabolites of UV-absorbing compounds, such as phenolic and flavonoids.

Therefore, irradiation intensity has influence on flavonoid metabolism, and vegetables exposed to full sunlight have demonstrated to contain more flavonois than those grown in the shade (Schreiner, 2005). The effect of UV-B radiation in stimulating the production of flavonoids and other phenolics could be explained by the induction of important enzymes of phenolic and flavonoid biosynthesis, such as phenylalanine ammonia-lyase (PAL), chalcone isomerase (CHS) and flavonol synthase (FLS) (Winkel-Shirly, 2002; Emilany et al., 2013).

Some studies have reported a positive effect of UV-B radiation on flavonoid production by several plant species (Schreiner et al., 2014) and in fruits and vegetables treated with UV-C (Arcas et al., 2000; Cantos et al., 2003). The enhancing effects on phenolic metabolism can be detected through quantitative and qualitative phenolic profiles in plant extracts (Neugart et al., 2012). Also, UV radiation, used in post-harvest as a sanitizing treatment, can induce biological stress in plants with the consequent production of phytoalexin compounds such as flavonoids or stilbenes. It was determined that resveratrol production responds positively to UV radiation and is considerably higher in leaves than in fruit (Douillet-Breuil et al., 1999). Analysis of grapes affected by *Botrytis cinerea* showed accumulation of resveratrol in areas surrounding fungal infection, suggesting a localized effect of the antimicrobial compound (Jeandet et al., 2002).

3.5. Cell Wall Yeast

Cell wall yeast can be considered an elicitor able to induce PAL and consequent accumulation of phytoalexins and other secondary metabolites in numerous plant species, including alfalfa, tobacco, lupines albus, apple, solanum khasianum and soybean (Muehlenbeck et al., 1996; Wojtaszek et al., 1997; Borejsza-Wysocki et al., 1999).

Yeast extracts contain several compounds that may act as elicitors. In this respect, yeast cell walls are made up of nanoproteins, β-1-3 and β-1-6 glucans and chitin, while yeast plasmatic membrane comprises lipids, sterols, and proteins (Kapteyn et al., 1999). Most of these compounds are regarded as triggers of various modes of plant defense (Ferrari, 2010). Several *in vitro* studies have reported the accumulation of secondary metabolites and the activation of PAL following yeast extract applications to plant cell cultures (Yan et al., 2006).

Shehata et al., (2012) found that in cucumbers, the treatment with active dry yeast increased plant growth and yield, among other parameters. Other study has shown that the exogenous application of a yeast extract to soybean increases the concentration of photosynthetic pigments, yield, phenolic content, and the antioxidant activity (Dawood et al., 2013).

3.6. Harpine

Harpine is a heat stable, glycine rich protein of bacterial origin. It has been described as an elicitor able to activate enzymes such as PAL from the polyphenol biosynthesis pathway. Several researchers found an increase in total phenols in several fruits, for instance, post-harvest treated peaches and jujbe (Danner et al., 2008).

Harpin has been applied as an effective postharvest treatment to prevent decay in oranges (Lucon et al., 2010), melons (Bi et al., 2010), apples (Dc Capdevile et al., 2003) and pears (Wang et al., 2006). In addition, field applications demonstrated its usefulness for controlling pathogen-borne diseases in passion fruits (Boro et al., 2011), pears, quince and loquat (Bastas et al., 2007).

3.7. Abscisic Acid (ABA)

Endogenous abscisic acid (ABA) is a well-known senescence-triggering plant hormone and plays important roles throughout plant life and development, affecting seed germination, plant growth, senescence and plant stress response. Especially, under some stresses like hyper-or hypo osmotic stress, salt, cold and drought stress. ABA also participates in the initiation of ripening and related changes in grape development (Wheeler et al., 2009).

It has been demonstrated that exogenous application of ABA increases the content of anthocyanins in grapes skins (Jeong et al., 2004) and can improve the colour and quality of the grapes (Cantin et al., 2007). Lacampagne et al., (2009) reported that ABA regulates enzyme involved in tannin biosynthesis and thus elevates tannin content in green grapes at veraison. These studies suggest that ABA plays a significant role in triggering the flavonoid biosynthetic pathway. Shandu et al., (2011) reported that ABA enhances the antioxidant capacity, anthocyanins and phenolic content of muscadine grapes but these effects may vary depending upon the cultivar and other environmental factors.

3.8. Others Elicitors

Many other substances have been studied as possible elicitors in different fruits and vegetables. For instance, ethephon is a widely used elicitor for metabolite production in plants, and in *Vitis vinifera*, this compound has been shown to increase anthocyanin content over two fold (Saw et al., 2012).

Relatively little information is available of K_2HPO_4 as defense elicitor in fruit crops. However, foliar sprays of phosphates have been shown to control different pathogens attacks, such as powdery mildew in field-grown nectarines, mango trees and grapevine (Reuveni and Reuveni, 1995) and also anthracnose disease in cashew (Lopez and Lucas, 2002).

Other compound named probenazole, the active component of *Oryzemate*, is mainly applied to rice crop to control rice blast, caused by a fungus Magnaporthe *grisea*, and bacterial leaf blight, caused by *Xanthomonas oryzae pv. oryzae*. It also appears to be effective in field trials on maize, protecting it against fungal leaf blight caused by *Cochliobolus heterostrophus* (Yang et al., 2011). Also in Arabidopsis, probenazole acts upstream of SA accumulation and leads to SAR mediated by SA, but not jasmonic acid or ethephon signaling (Yoshioka et al., 2001).

β-Aminobutyric acid (BABA), a non-protein amino acid, confers a broad-spectrum systemic resistance in different crop-pathogen systems (Gozzo and Faoro, 2013; Walters et al., 2013). BABA protective effect is due to a potentiation of defense mechanisms (Conrath, 2011). Besides this mechanism of defense, BABA also exhibits a direct fungitoxic effect on some pathogens such as *Leptosphaeria maculans* (Sasek et al., 2012).

Finally, treatment with phenylalanine (Phe) is also another possible approach used in influencing the biosynthetic pathway in plant cell system (Kovacik et al., 2007; Shinde et al., 2009). According to Fraser and Chapple (2011), Phe is an end product of shikimate pathway. Moreover, Phe served as an upstream metabolic precursor in the biosynthesis of phenolics and flavonoids through a series of enzymatic reactions, such as phenylalanine ammonia-lyase (PAL), cinnamic acid 4-hydroxylase (C4H) and 4-coumarate: CoA ligase (4CL) to yield 4-coumaroyl-CoA (Ritter and Schulz, 2004).

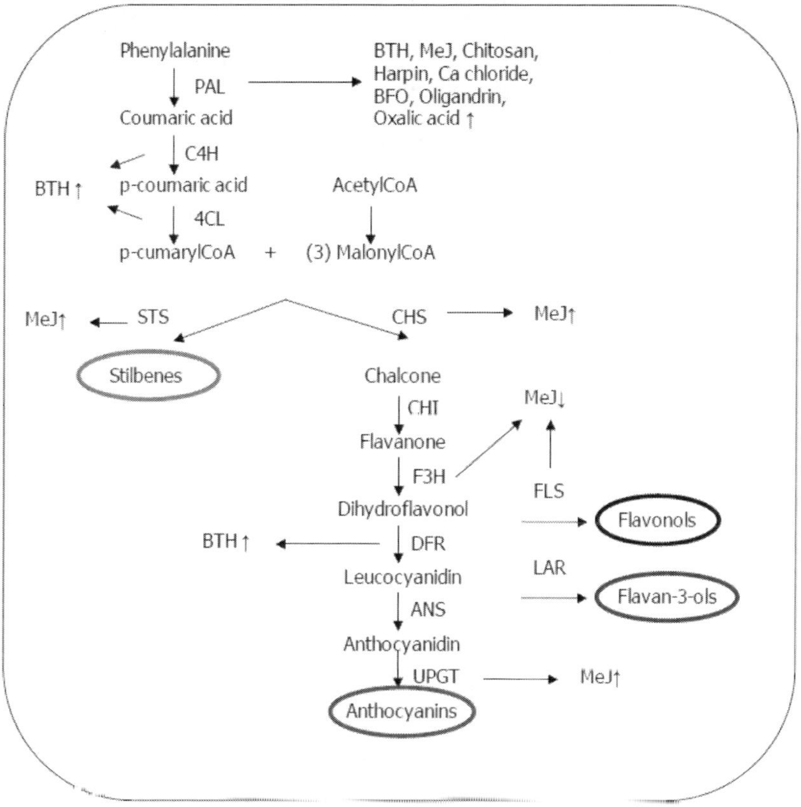

Abbreviations: 4-CL: 4-coumarate/coenzyme A ligase; ANS: anthocyanidin synthase; C4H: cinnamate 4-hydroxylase; CHI: chalcone isomerase; CHS: chalcone synthase; DFR: dihydroflavonol 4-reductase; F3'H: flavonoid 3'-hydroxylase; F3'5'H: flavonoid 3',5'-hydroxylase; FLS: flavonol synthase; LAR: leucoanthocyanidin reductase; PAL: phenylalanine ammonia-lyase; STS: stilbene synthase; UPGT: UDPglucose, flavonoid-*O*-transferase (UPGT); BTH: benzothiadiazole; MeJ: methyl jasmonate; BFO: Burdock fructooligosaccharide; ↑: increase in enzyme activity; ↓: decrease in enzyme activity.

Figure 3. Summary of polyphenols synthesis and enzyme regulation by elicitors (Ruiz-García and Gómez-Plaza, 2013).

4. Increasing Phenolic Compounds by Using Elicitors

Plants phenolics may be divided in two classes: (1) preformed phenolics that are syntetized during the normal development of plant in tissues and (2) induced phenolics that are synthetized by plants in response to physical injury, infection or when are stressed by suitable elicitors. The induction of phenylpropanoid metabolism can be achieved artificially by treatments with elicitors or exposure to specific stress conditions (Mater and Griming, 1994). Accumulation of such metabolites often occurs in plants treated with various elicitors, signal compounds, and abiotic stresses (Zhang et al., 2004).

Elicitors such as BTH, CHT, MeJ, JA, SA, β-aminobutyric acid, ozone, aluminium chloride, and UV-C light have been used to enhance nutraceutical grape properties (Cisneros-Zevallos, 2003; Fernández-Marin et al., 2012). Different works have shown the effect of elicitors over the increase in the biosynthesis of phenolic compounds (Table 3).

A lot of studies about this item have been done, so investigations about the effect of MeJ on berry composition have proven that this compounds affects the antioxidant capacity, quality parameters, aroma compounds and shelf-life of fruits (Blanch et al., 2011; Wang and Zheng, 2005). Chitosan increased the content of phenolic compounds in Greek oregano (Yin et al., 2011); sweet basil (Kim et al., 2006), butter lettuce (Zloteck and Jakubczyk, 2014) and apricot (Glasemezhad and Javaherdashti, 2008). Liu et al., (2013) reported that ethylene (100mg/L) improved total phenolic content and free radical scavenging-linked activity in mung bean sprouts. More recently, this research group found that kidney bean seeds treated with glutamic (5 mM), folic (50 µM) or ascorbic (500 µM) acid during germination for 8 days showed an enhanced content of total phenolic compounds and γ-aminobutyric acid as well as antioxidant (Limón et al., 2014). On the other hand, grape and wine are the main sources of resveratrol, and other stilbenes such as piceid, astringin, piceatannol, and viniferins, in diet. All these stilbenes exhibit a pronounced antioxidant activity, however, their concentration in grapes and wine are rather low (Guerrero et al., 2009).

Table 3. Phenolic compounds increased by elicitors

Plant Food	Elicitor Treatment	Application	Target Compounds Class and Increase	Reference
"Fuji" apples	Ethephon (2-chloroethyl phosphonic acid) (100 mg/L)	Sprayed for 4 weeks before commercial harvest	Anthocyanins (8-fold), and flavonols (2-fold) during fruit maturation	[90]
Strawberry fruits	CO_2 (ambient + 600 µmol)	28 months	Anthocyanin and flavonols (30%–50%)	[95]

In other works, synergistic effect on phytoalexin production has been described between MeJ and ethephon (Faurie et al., 2009), CHT and UV C (Rommanazzi et al., 2006), MeJ gas and UV-C (Larrondo et al., 2003), methyl jasmonate and cyclodextrins (Lijavetzky et al., 2008).

5. INCREASING FLAVONOLS COMPOUNDS BY USING ELICITORS

Flavonoids are known to play a protective role against microbial invasion in plants that synthesize these polyphenols in traditional medicine to treat infectious diseases. This protective role involves the presence of flavonoids in plants as constitutive agents as well as their accumulation as phytoalexins in response to microbial attack. Therefore, plants rich in flavonoids have been used for many years in traditional medicine to treat infectious diseases.

Flavonols are secondary metabolites present in almost all higher plants. They are considered to act as UV- and photo-protectors because they absorb strongly at both UV-A and UV-B wavelengths. Flavonols also play an important role in wine copigmentation together with anthocyanins, are useful markers in grape taxonomy, and are considered bioactive grape/wine compounds of possible importance for human health and nutrition. The chemical structure of flavonols is closely related with their biosynthesis. As all phenolic compounds, flavonols are products of the phenylpropanoid pathway, which converts phenylalanine into 4-coumaroyl-CoA and later into tetrahydroxychalcone (Figure 4). The biosynthesis of the various classes of flavonoids, which include flavonols, starts from this last metabolite. In general, flavonols are C6-C3-C6 polyphenolic compounds in which two hydroxylated benzene rings, A and B, are joined by a three-carbon chain which is part of a heterocyclic C ring with a 3-hydroxyflavone backbone, and a double bond. In detail, they differ in the number and type of substitution in the B ring.

Flavonols are mainly located in the outer epidermis of the skin, since they act as UV-protecting agents. Their synthesis begins in the flower buttons, and the highest concentrations are found a few weeks after veraison. It concentration is stable during early fruit development and decreases as the grape berries increase in size (Hermosín-Gutierrez et al., 2012). As regard to different kind of flavonols, in grapes and wines can be found mono (kaempferol), di (quercetin and isorhamnetin), and trihydroxylated (myricetin and syringetin) flavonol glycosides (glucosides and glucuronides and small quantities of galactosides), although these compounds are at a much lower concentration than anthocyanins. Flavonols are very close to anthocyanins in the biosynthetic pathway, indeed, they share most of the pathway, so that an increase in the activity of enzymes upstream in the flavonoid biosynthetic pathway may also affect the concentration of these compounds.

Kaempferol is a polyphenolic flavonoid and it is a natural plant product with potential pharmacological and nutraceutical activities. This compound is known for its health promoting effect and study for its medicinal and nutritional activities. This compound and its glycosides have been isolated from plants used in popular medicine for their antimicrobial properties. Numerous papers have reported that kaempferol, its glycosides, or plants containing kaempferol have antibacterial, antiviral, antifungal and antiprotozoal activities.

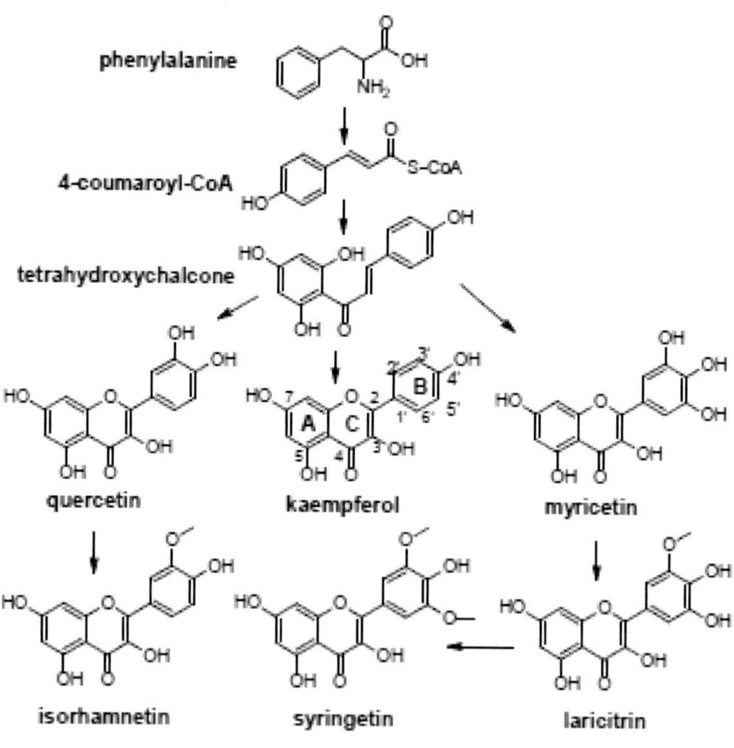

Figure 4. Flavonols pathway (Flamini et al., 2013).

On the other hand, quercetin is the most important flavonoid in experimental studies. It acts as antihistamine (which is useful in reducing allergy symptoms and help in reducing inflammation associated with various forms of arthritis). Quercetin also works as anti-inflammatory, antioxidant and anticancer substance. It helps in solving problems of cellular regeneration, hemorrhoids, menopausal symptoms and non-healing ulcers.

The total content and pattern of flavonols is highly variable across genotypes and can also be modulated to some extent by biotic and abiotic factors. Flavonol patterns are considered to be an important chemo-taxonomical parameter (Castillo-Muñoz et al., 2007). White and light red grape varieties synthesize mainly the mono- and di-substituted B-ring derivatives kaempferol, quercetin and isorhamnetin; red grapes also accumulate the tri-substitutes myricetin, laricitrin and syringetin (Hermosín-Gutierrez et al., 2012).

Quercetin is thus the major flavonol of all white varieties such as Chardonnay, Riesling, Viogner and Sauvignon Blanc, in which it represents over 70% of total flavonols. Quercetin is also the major flavonol in some light red/rosé varieties such as Nebbiolo, Pinot Noir, Sangiovese and Gewurztraminer, which also contain small percentages of myricetin (less than 20%). Conversely, myricetin is the major flavonol of most of the red varieties like Cabernet Sauvignon, Sagrantino and Teroldego (Mattivi et al., 2006). Another parameter which influences the amount of flavonols is the thickness of the berry skin, and thick-skinned grapes are reported to produce wines with higher amount of flavonols (e.g., Cabernet Sauvignon) than thin-skinned ones (e.g., Grenache) (McDonald et al., 1998).

Agronomic and environmental factors also strongly affect the amount and then the profile of flavonols in grape. In general, it would normally be expected that grapes more exposed to daylight could enhance the biosynthetic pathway of all flavonoids (Hermosin-Gutierrez et al., 2012). Some recent studies have shown that high temperatures during maturation decrease the expression of genes related to flavonoid synthesis and favor anthocyanin biosynthesis. Day temperatures of 15–25°C, falling to 10–20°C at night, produced grapes with higher amounts of flavonols with respect to higher daytime temperatures (30–35°C) (Kliewer et al., 1972). Azuma et al., (2012) recently demonstrated that total flavonol amounts were higher for a daytime temperature of 15°C under light treatment, with small variations in temperature, although in all cases the gene expression responsible for their biosynthesis was almost undetectable in bark-treated specimens. In particular, flavonol biosynthesis in plant tissues is greatly influenced by sunlight.

Therefore, elicitation is a strategy to enhance the production of flavonols in plants. These are only found in small quantity in grapes, which makes very expensive to produce them as commercial products (Castillo-Muñoz et al., 2007). Several studies have shown results about the increase of flavonols in plants by using elicitors. Ruiz-García et al., (2012) reported an increase in flavonols grapes in two years when BTH was applied to grapes and only in the second year when grapes were treated with MeJ. Wang and Zeng (2005) described an increase of flavonols in different fruits with the use of MeJ. Gozzo (2003) stated that the treatment with elicitors prepares the plant to react more efficiently (especially with regard to the activation of the phenyl propanoid pathway) when challenged with a pathogen, which could explain the greater increases in flavonols in grapes.

Although the influence of agrochemicals and plant activators on grape stilbenes, flavan-3-ols (Iriti et al., 2004) and anthocyanins is a known issue due to SAR (systemic acquired resistance) and activators induce the expression of phenylpropanoid genes in grapevine (Busam et al., 1997), little is reported about their influence on flavonols. An induction of their synthesis, similar to that observed for anthocyanins, could be expected, given the closeness of the two biosynthetic pathways. Iriti et al., (2004) revealed a very poor effect in the total flavonoids amount in Merlot variety during vintage 2004. However, this study was not specific for flavonols. In a two-year study with a treatment with BTH and methyl jasmonate (MeJ) conducted on Monastrell grape (2009–2010), an increase in flavonol concentration was observed (Ruiz-García et al., 2012) but the wide biological variability makes the results not completely consistent for the two years investigated. BTH-treated grapes had a higher flavonol concentration in both years (+17% and +56%), while the treatment with MeJ increased flavonols only in one year even if with a more pronounced effect (+131%). Differences between treated/untreated samples was more evident in the colder-humid year (2010), in which the conditions were more favorable for the development of pathogens, but no difference in the flavonol profile was observed. In any case, the influence of agrochemicals and plant activators in flavonoid synthesis is still an unclear argument in which further studies are needed.

Some works about the use of elicitors in different crops and the effect on the content of flavonols are shown in the next section.

5.1. Methyl Jasmonate

Moreno et al., (2010) reported in different fruits, an enhancement in the levels of myricetin, quercetin and kaempferol after postharvest treatments with MeJ. De la Peña-Moreno et al., (2010a, b), in recent investigations showed the effect of pre and post-harvest MeJ on certain flavonols in strawberries and red raspberries, observing decreases higher than 50% in quercetin and other flavonols after seven-day storage. It is hence suggested that MeJ promotes the activity of some enzyme, regulating the formation of flavonols and ellagic acid during ripening and postharvest period. Thus, preharvest MeJ increased flavonol content, whereas postharvest MeJ only allowed the natural drastic decline to be offset. Also, differences in pre- and postharvest MeJ effects are related with the maturation process of berries. Raspberry and blackcurrant are non-climacteric fruits and, therefore, once harvested, they never ripe further. During maturity flavonols and ellagic acid drop (Wang et al., 2009) and this decline becomes more drastic during storage. By the contrary Flores et al., (2015) showed an enhancement in myricetin, ellagic acid and quercetin content in berries treated with MeJ at preharvest and Wang et al., (2009) reported minor increases in myricetin content, although regarded as statistically significant by the authors, have been found in bayberries exposed to MeJ after harvesting.

In studies carried out by Portu et al., (2015) in Tempranillo grapes found that myricetin-type flavonols were predominant, followed by quercetin-type flavonols, accounting together for around 80% of total flavonol content. These studies showed that MeJ foliar application to Tempranillo grapevines has a stronger effect on anthocyanins than on flavonols. These authors also showed an improvement in the wine flavonol composition although the behavior was not the same in both years (Portu et al., 2016).

MeJ application may have a stronger effect in years when pathogen development is more suitable (Gozzo and Faoro, 2013; Ruiz-Garcia et al., 2012). The rain is an important factor in the results obtained. Ruiz-Garcia et al., (2012), observed that control wines and wines made from bunches treated with MeJ had similar flavonol content. Recent evidences suggest that MeJ, when applied to the leaves, may enhance the grape and wine quality by increasing the content of several phenolic compounds, including anthocyanins, stilbenes and to a lesser extent, flavonols (Portu et al., 2015).

5.2. UV-Radiation

Rodrigues et al., (2010) found that UV-radiation and ethylene application in two varieties of onions increased the flavonol content in 26% and 45%, respectively.

In a 2-year vineyard study, Spayd et al., (2002) imposed UV barriers over the canopy and fruiting zone and observed that flavonols (glucosides of quercetin, myricetin and kaempferol) all showed large significant increased with sun exposure. UV exposure is found to be significant in the accumulation of flavonols such as quercetin-3-glucoside. It was also noted that increased UV-B exposure dramatically increased the ratio of quercetin to kaempferol suggesting quercetin is a preferred metabolite for UV defense.

Dos Santos Nasciment et al., (2005) reported that the treatment with supplemental UV-B radiation changed the amount of flavonoids, increasing the total flavonoid and quercetin contents and also the phenolic profile of the *K. pinnata* extracts. Also, UV-B irradiation enhanced the concentration of flavonols in *Norway spruce* (*Picea abies*) and in silver birch and grape leaves (Tegelberg et al., 2004).

5.3. Chitosan (CHT)

Singh et al., (2016) found that in spinach leaves improved the following flavonols rutinosides: kaempferol, quercetin and isorhamnentin when spinaches were treated with CHT.

5.4. Benzothiadiazole (BTH)

Preharvest treatments with BTH in strawberries cultivated in greenhouses have proved to be useful for preventing powdery mildew and increasing the content of quercetin and kaempferol (Anttonen et al., 2003). This treatment also enhanced the accumulation of kaempferol hexose in leaves and kaempferol malonylglucoside in fruits (Hukkanen et al., 2007), increasing the amount of quercetin and kaempferol in berries (Karjalainen et al., 2002).

5.5. Abscisic Acid (ABA)

Fujita et al., (2006) reported an increase in the transcript levels of two flavonol synthase genes and an enhancement of quercetin content in the skin of Merlot grapes after veraison.

Shandu et al., (2011) investigated in Muscadine grapes the effect of ABA, when this elicitior was applied in two moments, and found that treated Noble grapes, a kind of Muscadine grapes cultivar, had higher concentrations of flavonols (myricetin, quercetin and kaempferol) compared to controls at first sampling. Similarly, the levels of myricetin, quercetin and kaempferol were enhanced by 54%, 45% and 48%, respectively, in treated Noble grapes compared to controls. A comparison of control groups from first and second sampling in Noble showed a significant increase in kaempferol content.

5.6. Cell Wall Yeasts

Portu et al., (2016) found only minor differences between the control wines and wines made with cell yeast extract treated grapes and chitosan treated grapes. These authors suggested that the application of elicitors at veraison usually exerts a limited impact on flavonol synthesis. Despite the fact of anthocyanins and flavonols share a big part of their metabolic pathway, it appears that anthocyanin biosynthesis is preferentially activated in comparison with flavonols. It must be taken into account that flavonols are important copigments that contribute to wine color stability.

5.7. Other Elicitors

Concentrations of leaf kaempferol increased under all sugar treatments up to a maximum of 40% in D- allose-treated plants. Both natural and rare hexose sugars are promising natural elicitors of physiological stress that can induce anticancer flavonoid kaempferol synthesis in soy plants (Cook et al., 2013).

Brassica species have been shown to contain quercetin and kaempferol as main flavonoid compounds, mainly occur as glycosides (quercetin 3-O-sophoroside and kaempferol 3-O-sophoroside) (Vallejo et al., 2004). Contrary, all studied extracts appeared to be very effective elicitors of kaempferol biosynthesis in sprouts (17-fold for 0.1% SC).

Dueñas et al., (2015) showed different results when different elicitors (ascorbic acid, folic acid and glutamic acid) were applied in kidney beans. Regarding flavonoid groups, germination brought about a general decrease in flavan-3-ols and anthocyanins that was compensated with a higher content of flavanones and flavonols. These changes in the phenolic composition could be explained by the activation of phenolic metabolism (Khandelwal et al., 2010) being dependent on legume type and germination conditions (López-Amorós et al., 2006). New phenolic compounds (eriodictyol-O-hexoside, quercetin glucuronide, kaempferol glucuronide and acylhexoside) were identified in kidney bean sprouts, indicating the activation of the phenylpropanoid pathway. This study showed that elicitors treatments during 8 days of germination promote the synthesis and accumulation of different flavonoid compounds (flavanones and flavonol-O-glycosides), depending on the elicitor type.

Previous *in vitro* and *in vivo* studies have demonstrated that certain flavonoids (anthocyanins, flavones, flavanones, flavanols and flavones) have an inhibitory effect on ACE activity, which plays a key role in the regulation of arterial blood pressure (Guerrero et al., 2012). Quercetin and kaempferol are among the most potent ACE inhibitors out of 39 phenolic compounds belonging to different structural subtypes (Al Shukor et al., 2013; Guerrero et al., 2012). Higher ACE inhibition activity of flavonols has been reported to be due to the combination of the following elements in the flavonoid structure: the catechol group in the B-ring, the double bond between C2 and C3 at the C-ring, and the cetone group in C4 at the C-ring (Guerrero et al., 2012).

Beccatti et al., (2014) showed a significant higher values in the following concentrations of myricetin, myricetin-3-glucoside, quercetin, and quercetin-3-glucuronide in wines elaborated with grapes treated in postharvest with ethylene when were compared with control wines.

In conclusion, elicitors can be a good alternative to increase flavonol content in different fruits and vegetables, due in first place, to their use as fungicides or pesticides environmental friend being able to defense crop plants against pests or diseases; and in a second place, due to their use as elicitor, depending on the doses applied, the application moment and the cultivar, can trigger the PAL dowmstream, and biosynthetize different flavonoids, among them, flavonols, a kind of phenolic compounds very important in human health.

ACKNOWLEDGMENTS

This work was made possible by financial assistance from the Instituto Nacional de Investigación y Tecnología Agraria y Alimentaria (RTA2013-00053-C03-02).

REFERENCES

Achakzai AKK, Achakzai P, Masood A, Kayani SA and Tareen RB (2009) Response of plant parts and age on the distribution of secondary metabolites on plants found in Quetta. *Pakistan Journal of Botany* 41:2129–2135.

Adebamowo CA, Cho E, Sampson L, Katan MB, Spiegelman D, Willett WC and Holmes MD (2005) Dietary flavonols and flavonol-rich foods intake and the risk of breast cancer. *International Journal of Cancer* 20:628-33.

Adikaram NKB (1990) Possibility of control of postharvest fungal diseases by manipulation of host defence systems. In: *Proceedings of the 3rd Conference on Plant Protection in the Tropics*, volume 5, Genting Highlands, Malaysia, pp. 31−36.

Al Shukor N, Van Camp J, Gonzales GB, Staljanssens D, Struijs K, Zotti MJ et al., (2013) Angiotensin-converting enzyme inhibitory effects by plant phenolic compounds: A study of structure activity relationships. *Journal of Agricultural and Food Chemistry* 61:11832–11839.

Anttonen M, Hukkanen A, Tiilikkala K and Karjalainen R (2003) Benzothiadiazole induces defense responses in berry crops. *Acta Horticulturae* 567:177–182.

Arcas MC, Botía JM, Ortuño AM and Del Río JA (2000) UV irradiation alters the levels of flavonoids involved in the defence mechanism of Citrus aurantium fruits against Penicillium digitatum. *European Journal of Plant Pathology* 106:617–622.

Ayala-Zavala JF, Wang SY, Wang CY and Gonzalez-Aguilar GA (2004) Methyl jasmonate in conjunction with ethanol treatment increases antioxidant capacity, volatile compounds and postharvest life of strawberry fruit. *LWT-Food Science and Technology* 37:687–695.

Azuma A, Yakushiji H, Koshita Y and Kobayashi S (2012) Flavonoid biosynthesis-related genes in grape skin are differentially regulated by temperature and light conditions. *Planta* 236:1067–1080.

Baenas N, Garcia-Viguera C and Moreno DA (2014) Biotic elicitors effectively increase the glucosinolates content in brassicaceae sprouts. *Journal of Agricultural and Food Chemistry* 62:1881–1889.

Baker LG, Specht CA, Donlin MJ and Lodge JK (2007) Chitosan, the deacetylated form of chitin, is necessary for cell wall integrity in *Cryptococcus neoformans. Eukaryotic Cell* 6:855–867.

Bassi D (2011) Effect of SAR inducer on grape secondary metabolites. Thesis Doctoral.

Bastas KK and Maden S (2007) Evaluation of host resistance inducers and conventional products for fire blight management in loquat and quince. *Phytoprotection* 88:93–101.

Becatti E, Genova G, Ranieri A and Tonutti P (2014) Postharvest treatments with ethylene on *Vitis vinifera* (cv Sangiovese) grapes affect berry metabolism and wine composition. *Food Chemistry* 159:257-66.

Belhadj A, Telef N, Saigne C, Cluzet S, Barrieu F, Hamdi S and Merillon JM (2008) Effect of methyl jasmonate in combination with carbohydrates on gene expression of PR proteins, stilbene and anthocyanin accumulation in grapevine cell cultures. *Plant Physiology and Biochemistry* 46:493–499.

Bellincontro A, Fardelli A, De Santis D, Botondi R and Mencarelli F (2006) Postharvest ethylene and 1-MCP treatments both affect phenols, anthocyanins, and aromatic quality of Aleatico grapes and wine. *Australian Journal of Grape and Wine Research* 12:141–149.

Benjakul S, Visessanguan W, Tanaka M, Ishizaki S, Suthidham R and Sungpech O (2000) Effect of chitin and chitosan on gelling properties of surimi from barred garfish (Hemiramphus far). *Journal of the Science of Food and Agriculture* 81:102–108.

Bent AF and Mackey D (2007) Elicitors, effectors, and R genes: the new paradigm and a lifetime supply of questions. *Annual Review of Phytopathology* 45:399-436.

Bi Y, Li Y and Ge Y (2007) Induced resistance in postharvest fruits and vegetables by chemicals and its mechanism. *Stewart Postharvest Review* 3:1-7.

Bi Y, Li YC, Ge YH and Wang Y (2010) Induced Resistance in melons by elicitors for the control of postharvest diseases. *Postharvest Pathology* 2:31–41.

Blanch GP, Flores G and Ruiz del Castillo ML (2011) Influence of methyl jasmonate in conjunction with ethanol on the formation of volatile compounds in berries belonging to the Rosaceae. *Postharvest Biology and Technology* 62:168–178.

Boller T and Felix G (2009) A renaissance of elicitors: Perception of microbe-associated molecular patterns and danger signals by pattern-recognition receptors. *Annual Review of Plant Biology* 60:379-406.

Borejsza-Wysocki W, Lester C, Attygalle AB and Hrazdina G (1999) Elicited cell suspension cultures of apple (*Malus x domestica*) cv. liberty produce biphenyl phytoalexins. *Phytochemistry* 50:231-235.

Boro MC, Beriam LOS and Guzzo SD (2011) Induced resistance against *Xanthomonas axonopodis* pv. passiflorae in passion fruit plants. *Tropical Plant Pathology* 36:74–80.

Botondi R, Lodola L and Mencarelli F (2011) Postharvest ethylene treatment affects berry dehydration, polyphenol and anthocyanin content by increasing the activity of cell wall enzymes in Aleatico wine grape. *European Food Research and Technology* 232:679–685.

Bueter CL, Specht CA and Levitz SM (2013) Innate sensing of chitin and chitosan. *PLoS Pathogen* 9:e1003080.

Burketova L, Trda L, Ott PG and Valentova O (2015) Bio-based resistance inducers for sustainable plant protection against pathogens. *Biotechnology Advances* 33:994-1004.

Burguieres E, McCue P, Kwon Y-I and Shetty K (2007) Effect of vitamin C and folic acid on seed vigour response and phenolic-linked antioxidant activity. *Bioresource Technology* 98:1393–1404.

Busam G, Junghanns KT, Kneusel RE, Kassemeyer H and Matern U (1997) Characterization and expression of caffeoyl-coenzyme A 3-*O*-methyltransferase proposed for the induced resistance response of *Vitis vinifera* L. *Plant Physiology* 115:1039–1048.

Cantín MC, Fidelibus MW and Crisosto CH (2007) Application of abscisic acid (ABA) at veraison advanced red color development and maintained postharvest quality of 'Crimson Seedless' grapes. *Postharvest Biology and Technology* 46:237-241.

Cantos E, Espín JC, Fernandez MJ, Oliva J and Tomás-Barberán FA (2003) Postharvest UV-C-irradiated grapes as a potential source for producing stilbene enriched red wines. *Journal of Agricultural and Food Chemistry* 51:1208–1214.

Cao J, Jiang W and He H (2005) Induced resistance in Yali pear (*Pyrus bretschneideri Rehd.*) fruit against infection by *Penicillium expansum* by postharvest infiltration of acibenzolar-S-methyl. *Journal of Phytopathology* 153:640– 646.

Castillo-Muñoz N, Gómez-Alonso S, García-Romero E and Hermosín-Gutiérrez I (2007) Flavonol profiles of *Vitis vinifera* red grapes and their single-cultivar wines. *Journal of Agricultural and Food Chemistry* 55:992–1002.

Cho MH, No HK and Prinyawiwatkul W (2008) Chitosan treatments affect growth and selected quality of sunflower sprouts. *Journal of Food Science* 73:S70–S77.

Cisneros-Zevallos L (2003) The use of controlled post-harvest abiotic stresses as a tool for enhancing the nutraceutical content and adding-value to fresh fruits and vegetables. *Journal of Food Science* 68:1560–1565.

Cook A, Onuaguluchi D, Jackson J, Patterson B, Gilbert L, Combe S, Ireland S, Muniruzzaman S, Boue S and Ratnayaka H (2013) Elicitation of physiological stress and kaempferol accumulation in soybean by sugar treatment. ASHS Annual Conference, Palm Dessert.

Conrath U (2011) Molecular aspects of defense priming. *Trends in Plant Science* 16:524–531.

Creelman R and Mullet JE (1997) Biosynthesis and action of jasmonates in plants. *Annual Review of Plant Physiology and Plant Molecular Biology* 48:355–381.

Croizier A, Jaganath IB and Clifford MN (2009) Dietary phenolics: chemistry, bioavility and effects on healths. *Natural Products Report* 26:1001-1043.

Croteau R, Kutchan TM and Lewis NG (2000) Natural products (secondary metabolites). In: *Biochemistry & Molecular Biology of Plants*; Buchanan, B., Gruissem, W., Jones, R., Eds.; ASPP: Rockville, MD, USA, pp. 1250–1318.

Danner MA, Sasso SAZ, Medeiros JGS, Marchese JA and Mazaro JM (2008) Indução de resistência à podridão-parda em pêssegos pelo uso de elicitores em pós-colheita. *Pesquisa Agropecuária Brasileira* 43:793-799.

Dawood MG, El-Lethy SR and Mervat S (2013) Role of methanol and yeast in improving growth, yield, nutritive value and antioxidants of soybean. *World Applied Sciences Journal* 26:6–14.

De Capdeville G, Beer SV, Watkins CB, Wilson CL, Tedeschi LO and Aist JR (2003) Pre- and post-harvest harpin treatments of apples induce resistance to blue mold. *Plant Disease* 87:39–44.

de la Peña Moreno F, Blanch GP, Flores G and Ruiz del Castillo ML (2010a) Impact of postharvest methyl jasmonate treatment on the volatile composition and flavonol content of strawberries. *Journal of the Science of Food and Agriculture* 90:989–994.

de la Peña Moreno F, Blanch GP and Ruiz del Castillo ML (2010b) (+)-Methyljasmonate-induced bioformation of myricetin, quercetina and kaempferol in red raspberries. *Journal of Agricultural and Food Chemistry* 58:11639–11644.

de Pascual-Teresa S, Moreno DA and García-Viguera C (2010) Flavanols and anthocyanins in cardiovascular health: a review of current evidence. *International Journal of Molecular Science* 11:1679-1703.

Dixon RA (2001) Natural products and plant disease resistance. *Nature* 411:843-847.

dos Santos Nasciment LB, Vinicius Leal-Costa M, Aragao Menezes E, Rodrigues Lopes V, Frazao Muzitano M, Soares Costa S and Schwartz Tavares E (2015) Ultraviolet-B radiation effects on phenolic profile and flavonoid content of Kalanchoe pinnata. *Journal of Photochemistry and Photobiology B: Biology* 148:73–81.

Douillet-Breuil AC, Jeandet P, Adrian M and Bessis R (1999) Changes in the phytoalexin content of various Vitis spp. in response to ultraviolet C elicitation. *Journal of Agricultural and Food Chemistry* 47:4456–4461.

Downey MO, Harvey JS and Robinson SP (2004) The effect of bunch shading on berry development and flavonoid accumulation in Shiraz grapes. *Australian Journal of Grape and Wine Research* 10:55–73.

Dueñas M, Martínez-Villaluenga C, Limón R, Peñas E and Frías J (2015) Effect of germination and elicitation on phenolic composition and bioactivity of kidney beans. *Food Research International* 70:55-63.

Ellis BE, Kuroki GW and Stafford HA (1994) Genetic engineering of plant secondary metabolism. Springer Science & Business Media.

Emiliani J, Grotewold E, Falcone Ferreyra L. and Casati P (2013) Flavonols protect Arabidopsis plants against UV-B deleterious effects. *Molecular Plant* 6:1376–1379.

Faurie B, Cluzet S, Corio-Costet MF and Merillon JM (2009) Methyl jasmonate/ethephon cotreatment synergistically induces stilbene production in *Vitis vinifera* cell suspensions but fails to trigger resistance to Erysiphe necator. *Journal International des Science de la Vigne et du Vin* 43:99–110.

Fernandez-Marın MI, Guerrero RF, Garcıa-Parrilla MC, Puertas B, Richard T, Rodriguez-Werner MA, Winterhalter P, Monti JP and Cantos-Villar E (2012) Isorhapontigenin: a novel bioactive stilbene from wine grapes. *Food Chemistry* 135:1353–1359.

Ferrari S (2010) Biological elicitors of plant secondary metabolites: Mode of action and use in the production of nutraceutic. In: M. T. Giardi, G. Rea, & B. Berra (Eds.), Bio-farms for nutraceuticals: Functional food and safety control by biosensors (pp. 152–166). New York: Landes Bioscience and Springer Science + Business Media LLC.

Fischer TC, Gosch C, Pfeiffer J, Halbwirth H, Halle C, Stich K and Forkmann G (2007) Flavonoid genes of pear (*Pyrus communis*). *Trees–Structure and Function* 21:521–529.

Flamini R, Mattivi F, Rosso MD, Arapitsas P and Bavaresco L (2013) Advanced knowledge of three important classes of grape phenolics: Anthocyanins, stilbenes and flavonols. *International Journal of Molecular Sciences* 14:19651-19669.

Flores G, Blanch GP and Ruiz del Castillo ML (2015) Postharvest treatment with (-) and (+) methyl jasmonate stimulates anthocyanin accumulation in grapes. *LWT-Food Science and Technology* 62:807-812.

Franceschi VR and Grimes HD (1991) Induction of soybean vegetative storage proteins and anthocyanins by low-level atmospheric methyl jasmonate. *Proceeding of the National Academy of Science* 88:6745–6749.

Fraser CM and Chapple C (2011) The phenylpropanoid pathway in Arabidopsis. The Arabidopsis Book 9 e0152, doi/10.1199/tab.0152.

Freepons D (1991) Chitosan, does it have a place in agriculture? *Proceedings of the Plant Growth Regulation Society of America* 11–19.

Friedrich L, Lawton K, Dincher S, Winter A, Staub T, Uknes S, Kessmann H and Ryals J (1996) Benzothiadiazole induces systemic acquired resistance in tobacco. *Plant Journal* 10:61-70.

Fujita M, Fujita Y, Noutoshi Y, Takahashi F, Narusaka Y, Yamaguchi-Shinozaki K and Shinozaki K (2006) Crosstalk between abiotic and biotic stress responses: a current view from the points of convergence in the stress signaling networks. *Current Opinion in Plant Biology* 9:436–442.

Garcia-Brugger AG, Lamotte O, Vandelle E, Bourque S, Lecourieux D, Poinssot B, Wendehenne D and Pugin A (2006) Early signaling events induced by elicitors of plant defenses. *Molecular Plant Microbe Interactions* 19:711-724.

Gates MA, Tworoger SS, Hecht JL, De Vivo I, Rosner B and Hankinson SE (2007) A prospective study of dietary flavonoid intake and incidence of epithelial ovarian cancer. *International Journal of Cancer* 121:2225-2232.

Ghasemnezhad M and Javaherdashti M (2008) Effect of methyl jasmonate treatment on antioxidant capacity, internal quality and postharvest life of raspberry fruit. *Caspian Journal of Environmental Sciences* 6:73–78.

Gonzalo-Diago A, Dizy M and Fernández-Zurbano P (2014) Contribution of low molecular weight phenols to bitter taste and mouthfeel properties in red wines. *Food Chemistry* 154:187-198.

Gorlach J, Volrath S, Knauf-Beiter G, Hengy G, Beckhove U, Kogel K-H, Oostendorp M, Staub T, Ward E, Kessmann H and Ryals J (1996) Benzothiadiazole, a novel class of inducers of systemic acquired resistance, activates gene expression and disease resistance in wheat. *Plant Cell* 8:629-643.

Gouvea DR, Gobbo-Neto L, Sakamoto HT, Lopes JLC (2012) Seasonal variation of the major secondary metabolites present in the extract of *Eremanthus mattogrossensis* less (Asteraceae: Vernonieae) leaves. *Química Nova* 35:2139–2145.

Gozzo F (2003) Systemic acquired resistance in crop protection: From nature to a chemical approach. *Journal of Agricultural and Food Chemistry* 51:4487–4503.

Gozzo F and Faoro F (2013) Systemic acquired resistance (50 years after discovery): moving from the lab to the field. *Journal of Agricultural and Food Chemistry* 61:12473–12491.

Guerrero RF, Garcıa–Parrilla MC, Puertas B, Cantos-Villar E (2009) Wine, resveratrol and health: a review. *Natural Products Communications* 4:635–658.

Guerrero L, Castillo J, Quiñones M, Garcia-Vallvé S, Arola L, Pujadas G et al., (2012) Inhibition of angiotensin-converting enzyme activity by flavonoids: Structure–activity relationship studies. *PLoS ONE* 7:e49493.

Hadwiger LA and Beckman JM (1980) Chitosan as a component of pea-Fusarium solani interactions. *Plant Physiology* 66:205–211.

Halsam E and Lilley TH (1988) Natural astringency in foodstuffs—A molecular interpretation. *Critical Reviews and Food Science and Nutrition* 27:1–40.

Hammerschmidt R, Métraux JP and Van Loon LC (2001) Inducing resistance: a summary of papers presented at the First International Symposium on Induced Resistance to Plant Diseases. Corfu, May 2000. *European Journal of Plant Pathology* 107:1−6.

Harborne JB (1993) Introduction to ecological biochemistry, 4th edition. London: Academic Press. 380 pp.

Harborne JB and Williams CA (2000) Advances in flavonoids research since 1992. *Phytochemistry* 55:481–504.

Harborne JB (2001) Twenty-five years of chemical ecology. *Natural Product Reports* 18:361-379.

Heredia JB and Cisneros-Zevallos L (2009) The effect of exogenous ethylene and methyl jasmonate on pal activity, phenolic profiles and antioxidant capacity of carrots (*Daucus carota*) under different wounding intensities. *Postharvest Biology and Technology* 51:242-249.

Hermosín-Gutiérrez I, Castillo-Muñoz N, Gómez-Alonso S and García-Romero E (2012) Flavonol Profiles for Grape and Wine Authentication. In: *Progress in Authentication of Food and Wine*. Ebeler, S.E., Takeoka, G. R., Winterhalter, P., Eds.; American Chemical Society: Washington, DC, USA; pp. 113–129.

Holopainen JK, Heijari J, Nerg AM, Vuorinen M and Kainulainen P (2009) Potential for the use of exogenous chemical elicitors in disease and insect pest management of conifer seedling production. *Open for Science Journal* 2:17-24.

Hukkanen AT, Kokko HI, Buchala AJ, Mcdougall GJ, Stewart D, Karenlampi SO and Karjalainen RO (2007) Benzothiadiazole induces the accumulation of phenolics and improves resistance to powdery mildew in strawberries. *Journal of Agricultural and Food Chemistry* 55:1862–1870.

Iriti M, Rossoni M, Borgo M, Ferrara L, Iriti M and Faoro F (2003) Constitutive and inducible immunity in plants. *Petria* 13:77-103.

Iriti M, Rossoni M, Borgo M and Faoro F (2004) Benzothiadiazole enhances resveratrol and anthocyanin biosynthesis in grapevine meanwhile inducing resistance to *Botrytis cinerea*. *Journal of Agricultural and Food Chemistry* 52:4406–4413.

Iriti M, Rossoni M, Borgo M, Ferrara L and Faoro F (2005) Induction of resistance to gray mold with Benzothiadiazole modifies amino acid profile and increases proanthocyanidins in grape: Primary versus Secondary Metabolism. *Journal of Agricultural and Food Chemistry* 53:9133-9139.

Iriti M, Vitalini S, Di Tommaso G, D'Amico S, Borgo M and Faoro F (2011) A new chitosan formulation induces grapevine resistance against powdery mildew and improves grape quality traits. *Australian Journal of Grape and Wine Research* 17:263-269.

Jansen MM, Hectors K, O'Brien NM, Guisez Y and Potters G (2008) Plant stress and human health: do human consumers benefit from UV-B acclimated crops? *Plant Science* 175:449–458.

Jeandet P, Douillet-Breuil AC, Bessis R, Debord S, Sbaghi M and Adrian M (2002) Phytoalexins from the Vitaceae: biosynthesis, phytoalexins gene expression in transgenic plants, antifungal activity and metabolism. *Journal of Agricultural and Food Chemistry* 50:2731–2741.

Jeong ST, Goto-Yamamoto N, Kobayashi S and Esaka M (2004) Effects of plant hormones and shading on the accumulation of anthocyanins and the expression of anthocyanin biosynthetic genes in grape berry skins. *Plant Science* 167:247–252.

Jing P, Zheng YH, Tang SS, Rui HJ and Wang CY (2009) Enhancing disease resistance in peach fruit with methyl jasmonate. *Journal of the Science of Food and Agriculture* 89:802–808.

Keinanen M, Oldham N and Baldwin IT (2001) Rapid HPLC screening of jasmonate-induced increases in tobacco alkaloids, phenolics, and diterpene glycosides in *Nicotiana attenuate*. *Journal of Agricultural of Food Chemistry* 49:3553-3558.

Khan AS and Singh Z (2007) Methyl jasmonate promotes fruit ripening and improves fruit quality in Japanese plum. *The Journal of Horticultural Science and Biotechnology* 82:695-706.

Kapteyn JC, Van Den Ende H and Klis FM (1999) The contribution of cell wall proteins to the organization of the yeast cell wall. *Biochimica et Biophysica Acta-General Subjects* 1426:373–383.

Karjalainen R, Lehtinen A, Keinanen M, Julkunen-Tiitto R, Hietaniemi V, Pihlava JM, Tiilikkala K and Jokinen K (2002) Benzothiadiazole and glycine betaine treatments enhance phenolic compound production in strawberry. *Acta Horticulturae* 567:353–356.

Khandelwal S, Udipi SA and Ghugre P (2010) Polyphenols and tannins in Indian pulses: Effect of soaking, germination and pressure cooking. *Food Research International* 43:526-530.

Kim H-J, Chen F, Wang X and Rajapakse NC (2006) Effect of methyl jasmonate on secondary metabolites of sweet basil (*Ocimum basilicum* L.). *Journal of Agricultural of Food Chemistry* 54:2327–2332.

Kliewer WM and Torres RE (1972) Effect of controlled day and night temperatures on grape coloration. *American Journal of Enology and Viticulture* 23:71–77.

Kovacik J, Kron I, Repcak M and Backor M (2007) Effect of feeding precursors on phenylalanine ammonia-lyase activity and coumarin accumulation in leaves of *Matricaria chamomilla* L. *Plant Growth Regulator* 52:9–15.

Kutchan TM (2001) Ecological arsenal and developmental dispatcher. The paradigm of secondary metabolism. *Plant Physiology* 125:58-60.

Lacampagne S, Gagne S and Gény L (2009) Involvement of abscisic acid in controlling the proanthocyanidin biosynthesis pathway in grape skin: new elements regarding the regulation of tannin composition and leucoanthocyanidin reductase (LAR) and anthocyanidin reductase (ANR) activities and expression. *Journal of Plant Growth Regulation* 29:81-90.

Lang G and Clausen T (1989) The use of chitosan in cosmetics. In: Skjak-Braek, G., Thorleif Anthosen, T., Standford, P. (Eds.), Chitin and Chitosan. Sources, Chemistry, Biochemistry. Physical Properties and Applications. Elsevier Applied Science, London and New York, pp. 139–147.

Lalel HJD, Singh Z and Tan SC (2003) The role of methyl jasmonate in mango ripening and biosynthesis of aroma volatile compounds. *The Journal of Horticultural Science and Biotechnology* 78:470–484.

Lara I (2013) Preharvest sprays and their effects on the postharvest quality of fruit. *Stewart Postharvest Review* 9:1-12.

Larrondo F, Gaudillere JP, Krisa S, Decendi A, Deffieux G and Merillon JM (2003) Airborne methyl jasmonate induces stilbene accumulation in leaves and berries of grapevine plants. *American Journal of Enology and Viticulture* 54:63–66.

Leifert W and Abeywardena M (2008) Grape seed and red wine polyphenol extracts inhibit cellular cholesterol uptake, cell proliferation, and 5-lipoxygenase activity. *Nutrition Research* 28:842–850.

Limón RI, Peñas E, Martínez-Villaluenga C and Frias J (2014) Role of elicitation on the health-promoting properties of kidney bean sprouts. *LWT-Food Science and Technology* 56:328–334.

Lin J, Rexrode KM, Hu F, Albert CM, Chae CU, Rimm EB, Stampfer MJ and Manson JE (2007) Dietary intakes of flavonols and flavones and coronary heart disease in US women. *American Journal of Epidemiology* 165:1305-1313.

Lin JH, Gong DQ, Zhu SJ, Zhang LJ and Zhang LB (2011) Expression of PPO and POD genes and contents of polyphenolic compounds in harvested mango fruits in relation to Benzothiadiazole-induced defense against anthracnose. *Scientia Horticulturae* 130:85–89.

Liu XD, Nishi N, Tokura S and Sakari N (2001) Chitosan coatedcotton fiber: preparation and physical properties. *Carbohydrate Polymere* 44:233–238.

Liu H, Jiang W, Bi B and Luo Y (2005) Postharvest BTH treatment induces resistance of peach (*Prunus persica* L. cv. Jiubao) fruit to infection by *Penicillium expansum* and enhances activity of fruit defense mechanisms. *Postharvest Biology and Technology* 35:263−269.

Liu L, Gu L, Ma Q, Zhu D and Huang X (2013) Resveratrol attenuates hydrogen peroxide-induced apoptosis in human umbilical vein endothelial cells. *European Review for Medical and Pharmacological Sciences* 17:88–94.

Lijavetzky D, Almagro L, Belchi-Navarro S, Martınez-Zapater JM, Bru R and Pedreño MA (2008) Synergistic effect of methyljasmonate and cyclodextrin on stilbene biosynthesis pathway gene expression and resveratrol production in Monastrell grapevine cell cultures. *BMC Research Notes* 1:132–140.

Lopez-Amoros ML, Hernandez T and Estrella I (2006) Effect of germination on legume phenolic compounds and their antioxidant activity. *Journal of Food Composition and Analysis* 19:277–283.

Lucon CMM, Guzzo SD, de Jesus CO and Pascholati SF (2010) Postharvest harpin or *Bacillus thuringiensis* treatments suppress citrus black spot in Valencia oranges. *Crop Protection* 29:766–772.

Martens S, Knott J, Seitz CA, Janvari L, Yu SN and Forkmann G (2003) Impact of biochemicalpre-studies on specific metabolic engineering strategies of flavonoid biosynthesis in plant tissues. *Biochemical Engineering Journal* 14:227–235.

Martin D, Tholl D, Gershenzon J and Bohlmann J (2002) Methyl jasmonate induces traumatic resin ducts, terpenoid resin biosynthesis and terpenoid accumulation in developing xylemof Norway spruce (*Picea abies*) stems. *Plant Physiology* 129:1003–1018.

Matern U and Grimmig B (1994) Natural phenols as stress metabolites. *Acta Horticulturae* 381:448–462.

Mattivi F, Guzzon R, Vrhovsek U, Stefanini M and Velasco R (2006) Metabolite profiling of grape: Flavonols and anthocyanins. *Journal of Agricultural and Food Chemistry* 54:7692–7702.

Mazaro SM, Deschamps C, de Mio LL, Biasi LA, de Gouvea A and Sautter CK (2012) Post harvest behavior of strawberry fruits after pre harvest treatment with chitosan and acibenzolar-*S*-methyl. *Revista Brasileira de Fruticultura* 30.185–190.

McDonald MS, Hughes M, Burns J, Lean MEJ, Matthews D and Crozier A (1998) Survey of the free and conjugated myricetin and quercetin content of red wines of different geographical origins. *Journal of Agricultural and Food Chemistry* 46:368–375.

Meng XH, Qin GZ and Tian SP 2010. Influences of preharvest spraying *Cryptococcus laurentii* combined with postharvest chitosan coating on postharvest diseases and quality of table grapes in storage. *Food Science and Technology* 43:596–601.

Meng XC, Tang YX, Zhang AY, Huang XM and Zhang ZQ (2012) Effect of oligochitosan on development of *Colletotrichum musae in vitro* and *in situ* and its role in protection of banana fruits. *Fruits* 67:147–155.

Moreno FD, Blanch GP and del Castillo MLR (2010) (+)-Methyl jasmonate-induced bioformation of myricetin, quercetin and kaempferol in red raspberries. *Journal of Agricultural and Food Chemistry* 58:11639–11644.

Muehlenbeck U, Kortenbusch A and Barz W (1996) Formation of hydroxycinnamoylamides and a-hydroxyacetovanillone in cell cultures of *Solanum khasianum. Phytochemistry* 42:1573-1579.

Myoshi H, Shimura K, Watanabe K and Onodera K (1992) Characterization of some fungal chitosans. *Biosciences, Biotechnology and Biochemistry* 56:1902-1905.

Naghiloo S, Movafeghi A, Delazar A, Nazemiyeh H, Asnaashari S and Reza Dadpour M (2012) Ontogenetic variation of total phenolics and antioxidant activity in roots, leaves and flowers of *Astragalus compactus Lam. (Fabaceae). Bioimpacts* 2:105-109.

Namdeo AG (2007) Plant cell elicitation for production of secondary metabolites. A review. *Pharmacognosy Reviews* 1:69-79.

Nandeeshkumar P, Sudisha J, Ramachandra KK, Prakash HS, Niranjana SR and Shekar SH (2008) Chitosan induced resistance to downy mildew in sunflower caused by *Plasmopara halstedii. Physiological and Molecular Plant Pathology* 72:188-194.

Ncube B, Finnie JF, Van J and Staden J (2012) Quality from the field: the impact of environmental factors as quality determinants in medicinal plants. *South African Journal of Botany* 82:11–20.

Neugart S, Klaring HP, Zietz M, Schreiner M, Rohn S, Kroh LW and Krumbein A (2012) The effect of temperature and radiation on flavonol aglycones and flavonol glycosides of kale (*Brassica oleracea var. sabellica*). *Food Chemistry* 133:1456–1465.

Noble AC (1994) Bitterness in wine. *Physiology and Behaviour* 56:1251–1255.

Odjacova M and Hadjiivansva C (2001) The complexity of pathogen defense in plants. *Bulgarian Journal of Plant Physiology* 27:101-109.

Oksman-Caldentey KM and Inze D (2004) Plant cell factories in the post-genomic era: new ways to produce designer secondary metabolites. *Trends in Plant Science* 9:433-440.

Ozeretskovskaya OL and Vasyukova NI (2002) The use of elicitors for protection of cultured plants demands caution. *Applied Biochemistry and Microbiology* 38:277-279.

Parr AJ and Bolwell GP (2000) Phenols in the plant and in man. The potential for possible nutritional enhancement of the diet by modifying the phenols content or profile. *Journal of the Science of Food and Agriculture* 80:985-1012.

Pasini F, Verardo V, Cerretani L, Caboni MF and D'Antuono LF (2011) Rocket salad (*Diplotaxis and Eruca spp.*) sensory analysis and relation with glucosinolate and phenolic content. *Journal of the Science of Food and Agriculture* 91:2858–2864.

Pena-Cortes H, Barrios P, Dorta F, Polanco V, Sanchez C, Anchez E and Ramirez I (2005). Involvement of jasmonic acid and derivatives in plant responses to pathogens and insects and in fruit ripening. *Journal Plant Growth Regulation* 23:246–260.

Percival D and MacKenzie JL (2007) Use of plant growth regulators to increase polyphenolic compounds in the wild blueberry. *Canadian Journal of Plant Science* 87:333–336.

Perez-Lamela C, Garcia-Falcon MS, Simal-Gandara J and Orriols-Fernandez I (2007) Influence of grape variety, vine system and enological treatments on the colour stability of young red wines. *Food Chemistry* 101:601-606.

Pezzuto JM (2008) Grapes and human health: A perspective. *Journal of Agricultural and Food Chemistry* 56:6777-6784.

Portu J, Santamaría P, López-Alfaro I, López R and Garde-Cerdán T (2015) Methyl jasmonate foliar application to Tempranillo vineyard improved grape and wine phenolic content. *Journal of Agricultural and Food Chemistry* 63:2328–2337.

Portu, J, López R, Baroja E, Santamaría P and Garde-Cerdán T (2016) Improvement of grape and wine phenolic content by foliar application to grapevine of three different elicitors: Methyl jasmonate, chitosan, and yeast extract. *Food Chemistry* 201:213-220.

Repka V (2001) Elicitor-stimulated induction of defense mechanisms and defense gene activation in grapevine cell suspension cultures. *Biologia Plantarum* 44:555–565.

Reglinski T, Dann E and Deverall B (2007) Integration of induced resistance in crop protection. *In*: Induced resistance for plant defence. A sustainable approach to crop protection. D. Walters, A. Newton, G. Lyon (Eds). Blackwell Publishing, Oxford, UK, pp. 201-228.

Ren H, Endo H and Hayashi T (2001) Antioxidative and antimutagenic and polyphenol content of pesticide-free and organically cultivated green vegetable using water-soluble chitosan as a soil modifier and leaf surface spray. *Journal of the Science of Food and Agriculture* 81:1426–1432.

Reuveni M and Reuveni R (1995) Efficacy of foliar application of phosphates in controlling powdery mildew fungus on field grown wine grapes: effect on cluster yield and peroxidase activity in berries. *Journal of Phytopathology* 143:21–25.

Ritter H and Schulz GE (2004) Structural basis for the entrance into the phenylpropanoid metabolism catalyzed by phenylalanine ammonia-lyase. *Plant Cell* 16:3426–3436.

Roberts MF and Wink M (1998) Alkaloids – biochemistry, ecological functions and medical applications. New York: Plenum 486 pp.

Rodriguez RE, Mecchia MA, Debernardi JM, Schommer C, Weigel D and Palatnik JF (2010) Control of cell proliferation in *Arabidopsis* thaliana by microRNA miR396. *Development* 137:103–112.

Romanazzi G, Nigro E, Ippolito A and Di Venere D (2002) Effects of pre- and postharvest chitosan treatments to control storage grey mold of table grapes. *Journal of Food Science* 67:1862–1867.

Romanazzi G, Gabler FM and Smilanick JL (2006) Preharvest chitosan and postharvest UV irradiation treatments suppress gray mold of table grapes. *Plant Disease* 90:445–450.

Rudell DR, Fellmann JK and Mattheis JP (2005) Preharvest application of methyl jasmonate to 'Fuji' apples enhances red coloration and affects fruit size, splitting, and bitter pit incidence. *Hortscience* 40:1760–1762.

Ruggiero A, Vitalini S, Burlini N, Bernasconi S and Iriti M (2013) Phytosterol in grapes and wine, and effects of agrochemicals on their levels. *Food Chemistry* 141:3473–3479.

Ruiz-Garcia Y, Romero Cascales I, Gil-Munoz R, Fernandez-Fernandez JI, Lopez- Roca JM and Gomez-Plaza E (2012) Improving grape phenolic content and wine chromatic characteristics through the use of two different elicitors: methyl jasmonate versus benzothiadiazole. *Journal of Agricultural and Food Chemistry* 60:1283–1290.

Ruiz-García Y and Gómez-Plaza E (2013) Elicitors: A tool for improving phenolic compounds. *Agriculture* 3:33-52.

Ryals JA, Neuenschwander UH, Willits MG, Molina A, Steiner HY and Hunt MD (1996) Systemic acquired resistance. *Plant Cell* 8:1809–1819.

Šašek V, Nováková M, Dobrev M, Valentová PI and Burketová L (2012) β-aminobutyric acid protects Brassica napus plants from infection by Leptosphaeria maculans. Resistance induction or a direct antifungal effect? *European Journal of Plant Pathology* 133:279–289.

Sandford P (1989) Chitosan: commercial uses and potential applications. In: Skjak-Braek, G., Anthosen, T., Standford, P. (Eds.), Chitin and Chitosan. Sources, Chemistry, Biochemistry. Physical Properties and Applications. Elsevier Applied Science, London and New York, pp. 51–69.

Saniewski M, Miyamoto K and Ueda J (1998) Methyl jasmonate induces gums and stimulates anthocyanin accumulation in peach shoots. *Journal of Plant Growth Regulation* 17:121–124.

Santos-Buelga C and Scalbert A (2000) Proanthocyanidins and tannin like compounds– nature, occurrence, dietary intake and effects on nutrition and health. *Journal of the Science of Food and Agriculture* 80:1094–1117.

Saw N, Riedel H, Cai Z, Kütük O and Smetanska I (2012) Stimulation of anthocyanin synthesis in grape (*Vitis vinifera*) cell cultures by pulsed electric fields and ethephon. *Journal of Plant Biotechnology* 108:47-54.

Scarponi L, Buonaurio R and Martinetti L (2001) Persistence and translocation of a benzothiadiazole derivative in tomato plants in relation to systemic acquired resistance against *Pseudomonas syringae* pv. *tomato*. *Pest Management Science* 57:262–268.

Schijlen E, de Vos CHR, Jonker H, van den Broeck H, Molthoff J, van Tunen A, Martens S and Bovy A (2006) Pathway engineering for healthy phytochemicals leading to the production of novel flavonoids in tomato fruit. *Plant Biotechnology Journal* 4:433–444.

Schreiner M (2005) Vegetable crop management strategies to increase the quantity of phytochemicals. *European Journal of Nutrition* 44:85-94.

Schreiner M, Martínez-Abaigar J, Glaab J and Jansen M (2014) UV-B induced secondary plant metabolites. *Optics and Photonics* 9:34–37.

Schurter R, Kunz W and Nyfeler R (1987) Process and a composition for immunizing plants against diseases. US Patent 4931581.

Sandhu AK and Gu L (2010) Antioxidant capacity, phenolic content, and profiling of phenolic compounds in the seeds, skin, and pulp of Vitis rotundifolia (muscadine grapes) as determined by HPLC-DAD-ESI-MS(n). *Journal of Agricultural and Food Chemistry* 58:4681–4692.

Sandhu AK, Gray DJ, Lu J and Gu L (2011) Effects of exogenous abscisic acid on antioxidant capacities, anthocyanins, and flavonol contents of muscadine grape (*Vitis rotundifolia*) skins. *Food Chemistry* 126:982–988.

Sharma P, Jha AB, Dubey RS and Pessarakli M (2012) Reactive oxygen species, oxidative damage, and antioxidative defense mechanism in plants under stressful conditions. *Journal of Botany* 1–26.

Shafiq M, Singh Z and Khan AS (2011) Pre-harvest spray application of methyl jasmonate improves red blush and flavonoid content in 'Cripps Pink' apple. *The Journal of Horticulural Science and Biotechnoly* 86:422–430.

Shafiq M, Singh Z and Khan AS (2013) Time of methyl jasmonate application influences the development of 'Cripps Pink' apple fruit colour. *Journal of the Science of Food and Agriculture* 93:611–618.

Shehata SA, Fawzy ZF and El-Ramady HR (2012) Response of cucumber plants to foliar application of chitosan and yeast under greenhouse conditions. *Australian Journal of Basic and Applied Sciences* 6:63–71.

Shinde AN, Malpathak N and Fulzele DP (2009) Optimized production of isoflavones in cell cultures of *Psoralea corylifolia L.* using elicitation and precursor feeding. *Biotechnology and Bioprocess Engineering* 4:612–618.

Singh G, Gavrieli J, Oakey JS and Curtis WR (1998) Interaction of methyl jasmonate, wounding and fungal elicitation during sesquiterpene induction in *Hyoscyamus muticus* in root cultures. *Plant Cell Reports* 17:391-395.

Singh S (2016) Enhancing phytochemical levels, enzymatic and antioxidant activity of spinach leaves by chitosan treatment and an insight into the metabolic pathway using DART-MS technique. *Food Chemistry* 199:176-184.

Siddiqui ZS and Zaman AU (2004) Effects of benlate systemic fungicide on seed germination, seedling growth, biomass and phenolic contents in two cultivars of *Zea mays* L. *Pakistan Journal of Botany* 36:577-582.

Spayd SE, Tarara JM, Mee DL and Ferguson JC (2002) Separation of sunlight and temperature effects on the composition of *Vitis vinifera* cv. Merlot berries. *American Journal of Enology and Viticulture* 53:171–182.

Stobiecki M, Matysiak-Kata I, Franski R, Skała J and Szopa J (2002) Monitoring changes in anthocyanin and steroid alkaloidglycoside content in lines of transgenic potato plants using liquidchromatography/mass spectrometry. *Phytochemistry* 62:959–969.

Swain T (1977) Secondary compounds as protective agents. *Annual Reviews of Plant Physiology* 28:479-501.

Taiz L and Zeiger E (2006) Plant Physiology. Sinauer Associates Inc. Publishers, Sunderland, Massachusetts, 4, 312-315.

Tassoni A, Fornale S, Franceschetti M, Musiani F, Micheal AJ, Perry B and Bagni N (2005) Jasmonates and Na-orthovanadate promote resveratrol production in *Vitis vinifera L. cv.* Barbera cell cultures. *New Phytologist* 166:895-905.

Tegeber R, Julkenen-Titto R and Aphalo PJ (2004) Red: Fared light ratio and UV-B radiation: Their effects on leaf phenolic and grown of silver birch seeding. *Plant All Enviroment* 27:1005-1013.

Terry LA and Joyce DC (2004) Elicitors of induced disease resistance in postharvest horticultural crops: a brief review. *Postharvest Biology and Technology* 32:1-13.

Tian S, Wan Y, Qin G and Xu Y (2006) Induction of defense responses against *Alternaria* rot by different elicitors in harvested pear fruit. *Applied Microbiology and Biotechnology* 70:729–734.

Tsiantos J, Psallidas P and Chatzaki A (2003) Efficacy of alternatives to antibiotic chemicals for the control of fire blight of pears. *Annals of Applied Biology* 143:319 –323.

Tzortzakis NG (2007) Methyl jasmonate-induced suppression of anthracnose rot in tomato fruit. *Crop Protection* 26:1507–1513.

Vallejo F, Tomás-Barberán FA and Ferreres F (2004) Characterization of flavonols in broccoli (*Brassica oleracea* L. var. *italica*) by liquid chromatography-UV diode-array detection-electrospray ionisation mass spectrometry. *Journal of Chromatography A* 1054:181–193.

Verna N and Shukla S (2015) Impact of various factors responsible for fluctuation in plant secondary metabolites. *Journal of Applied Research on Medicinal and Aromatic Plants* 2:105-113.

Veersham C (2004) In Elicitation: Madicinal Plant Biotechnology, CBS Publisher, India, 270-293.

Verpoorte R (1998) Exploration of nature's chemodiversity: the role of secondary metabolites as leads in drug development. *Drug Discovery Today* 3:232-238.

Walters DR and Fountaine JM (2009) Practical application of induced resistance to plant diseases: an appraisal of effectiveness under field conditions. *Journal of Agricultural Science* 147:523-535.

Walters DR, Ratsep J and Havis ND (2013) Controlling crop diseases using induced resistance: challenges for the future. *Journal of Experimental Botany* 64:1263–1280.

Wang SY and Zheng W (2005) Preharvest application of methyl jasmonate increases fruit quality and antioxidant capacity in raspberries. *International Journal of Food Science* 40:187–195.

Wang SY, Bowman L and Ding M (2008) Methyl jasmonate enhances antioxidant activity and flavonoid content in blackberries (*Rubus sp.*) and promotes antiproliferation of human cancer cells. *Food Chemistry* 107:1261–1269.

Wang KT, Jin P, Cao SF, Shang HT, Yang ZF and Zheng YH (2009) Methyl Jasmonate Reduces Decay and Enhances Antioxidant Capacity in Chinese Bayberries. *Journal of Agricultural and Food Chemistry* 57:5809–5815.

Wheeler S, Loveys B, Ford C and Davies C (2009) The relationship between the expression of abscisic acid biosynthesis genes, accumulation of abscisic acid and the promotion of *Vitis vinifera* L. berry ripening by abscisic acid. *Australian Journal of Grape and Wine Research* 15:195–204.

Wink M (1988) Plant breeding: importance of plant secondary metabolites for protection against pathogens and herbivores. *Theorical and Applied Genetics* 75:225–233.

Wink M (2006) Importance of plant secondary metabolites for protection against insects and microbial Infections. Rai and Carpinella (eds.) Naturally Occurring Bioactive Compounds, chapter 11, pp: 251-268.

Winkel-Shirley B (2002) Biosynthesis of flavonoids and effects of stress. *Current Opinion in Plant Biology* 5:218–233.

Wojtaszek P, Stobiecki M and Bolwell GP (1997) Changes in the composition of exocellular proteins of suspension-cultured Lupinus albus cells in response to fungal elicitors or CuCl2. *Journal of Experimental Botany* 48:2015-2021.

Yan Q, Shi M, Ng J and Wu JY (2006) Elicitor-induced rosmarinic acid accumulation and secondary metabolism enzyme activities in Salvia miltiorrhizahairy roots. *Plant Science* 170:853–858.

Yang B, Yongcai L and Yonghong G (2007) Induced resistance in postharvest fruits and vegetables by chemicals and its mechanism. *Stewart Postharvest Review* 3:1−7.

Yang SY, Chen YL, Feng LY, Yang E, Su XG and Jiang YM (2011) Effect of Methyl jasmonate on pericarp browning of postharvest lychees. *Journal of Food Processing and Preservation* 35:417–422.

Yao H and Tian S (2005) Effects of pre-and post-harvest application of salicylic acid or methyl jasmonate on inducing disease resistance of sweet cherry fruit in storage. *Postharvest Biology and Technology* 35:253–262.

Yin H, Fretté XC, Christensen LP and Grevsen K (2011) Chitosan oligosaccharides promote the content of polyphenols in Greek oregano (*Origanum vulgare* ssp. *hirtum*). *Journal of Agricultural and Food Chemistry* 60:136–143.

Yoshioka K, Nakashita H, Klessig DF and Yamaguchi I (2001) Probenazole induces systemic acquired resistance in *Arabidopsis* with a novel type of action. *Plant Journal* 25:149–157.

Yu O, Shi J, Hession AO, Maxwell CA, McGonigle B, Odell ZC, Yan Q, Cheuk W and Wu J (2003) Enhancement of tanshinone production in

Salvia miltiorrhiza hairy root culture by Ag+ elicitation and nutrient feeding. *Planta Medica* 70:147–151.

Zhang ZG, Wang YC, Li J, Ji R, Shen G, Wang SC, Zhou X and Zheng XB (2004) The role of SA in hypersensitive response and systemic acquired resistance induced by the elicitor PB90 of Phytophthora boehmeriae. *Physiological and Molecular Plant Pathology* 65:31–38.

Zeng K, Cao J and Jiang W (2006) Enhancing disease resistance in harvested mango (*Mangifera indica* l. cv. 'Matisu') fruit by salicylic acid. *Journal of the Science of Food and Agriculture* 86:694–698.

Zhao J, Davis LC and Verporte R (2005) Elicitor signal transduction leading to production of plant secondary metabolites. *Biotechnology Advances* 23:283-333.

Zhu SJ and Ma BC (2007) Benzothiadiazole- or methyl jasmonate-induced resistance to *Colletotrichum musae* in harvested banana fruit is related to elevated defense enzyme activities. *Journal of Horticultural Sciences and Biotechnoly* 82:500–506.

Zhu SJ, Zhang ZW, Xu JW, Ma LY and Tang WL (2007) Effect of BTH treatment on storability and activity of related enzymes of harvested loquat fruit. *Acta Horticulturae* 750:445–450.

Złotek U, Świeca M and Jakubczyk A (2014) Effect of abiotic elicitation on main health-promoting compounds, antioxidant activity and commercial quality of butter lettuce (*Lactuca sativa* L.). *Food Chemistry* 148:253–260.

In: Kaempferol ISBN: 978-1-63485-828-1
Editors: T.Garde-Cerdán and A.G. Diago ©2016 Nova Science Publishers, Inc.

Chapter 3

Therapeutic Uses of Kaempferol: Anticancer and Antiinflammatory Activity

Ignacio Gutiérrez-del-Río, Claudio J. Villar and Felipe Lombó[*]
Research Group Biotechnology of Nutraceuticals and Bioactive Compounds (BIONUC), Area Microbiology, University Institute of Oncology of Asturias (IUOPA), University of Oviedo, Oviedo, Spain

Abstract

Strong epidemiological evidences show an inverse relationship between frequent consumption of fruits and vegetables and the incidence of some types of health disorders, including certain cancers like colon, lung, ovarian, gastric or pancreatic cancer. Recent investigations have identified several diet phytochemicals that could be involved in cancer prevention, as for example flavonols like kaempferol.

In spite of recent progress in cancer treatment, classic chemotherapy has got numerous limitations like a low selectivity and tumor recurrence due to resistance development. These circumstances make necessary the search of new chemopreventive or chemotherapeutic agents. Kaempferol is gaining importance as phytochemical because of its wide range of

---

[*] Corresponding Author address: Email: lombofelipe@uniovi.es.

pharmacological properties. The anticancer effect of this flavonoid is mediated through different modes of action including antiproliferative activity, apoptosis induction, cell-cycle arrest, antiangiogenic and antimetastasic activities.

Kaempferol shows a demonstrated antiproliferative activity in different *in vitro* studies, showing apoptosis induction in different human cancer cell lines like leukemia, prostate, ovarian, oral cavity or colorectal cancer. This flavonol is able to induce cell-cycle arrest in HT-29 and Caco-2 human colorectal cancer cell lines by inhibiting DNA biosynthesis, thus leading to a downregulation in expression of diverse cyclines (D1, E and A), an also of CDK2 and CDK4; giving rise to a suppression in RB phosphorylation and cell cycle arrest in G1 phase. Besides, other studies made in the same colon cancer cell lines showed that kaempferol downregulates PI3K/Akt and ERK-1/2 pathways activation by inhibiting IGF-IR and ErbB3. Other studies made in OVCAR-3 cancer cell line (human ovarian cancer) demonstrated angiogenesis inhibition both by HIF-dependent pathway (Akt/HIF) and HIF-independent pathway (ESRRA). Kaempferol plays also an important role in the invasion capacity of human breast cancer cells by downregulating metalloproteinase-9 expression and activity. A noteworthy activity of kaempferol is its inhibition of Abcg2 transporters (multidrug transporter, MDR) in human breast cancer cell lines. Finally, kaempferol can also inhibit histone deacetylases (HDACs).

This flavonol is also able to supress inflammatory responses *in vitro* in an effective way, not only by its ROS scavenging activity, but also by its ability to directly inhibit Src, Syk, IRAK1 and IRAK2, which play a crucial role in NF-kB and AP-1 activation. These mechanisms are important as chronic inflammation and cancer are tightly connected processes.

However, kaempferol efficacy against *in vivo* tumor development and progression is impaired by its low bioavailability. One solution to this problem is the use of nanopharmaceuticals. Phytochemicals encapsulation in nanocarriers can improve their bioavailability, enhancing its specific cellular internalization by tumor cells, reducing phytochemical doses, and allowing the development of combined therapies including phytochemicals together with classical chemotherapy compounds.

1. ANTITUMOR ACTIVITY OF KAEMPFEROL

1.1. Natural Compounds in Cancer Chemoprevention and Treatment

'Cancer' is a generic term that encompasses a wide range of diseases affecting different body areas. It is generated by a multi-step and multi-mechanism process called oncogenesis that consists of three phases: initiation, promotion and progression (Nakamura et al., 2005). A defining characteristic of this disease is the rapid creation of abnormal cells that grow beyond their usual boundaries and invade adjacent areas, spreading to other organs through a process called metastasis, which is the final cause of cancer death. Cancer ranks as one of the leading causes of morbidity and mortality worldwide, with approximately 14 million new cases and 8.2 million deaths in 2012. It is also expected that the number of cases will increase about 70% over the next two decades reaching 22 million annual cases (Rajendran et al., 2014; WHO, 2014).

The search for a perfect antitumor drug is therefore urgent, and an ideal anticancer chemotherapeutic agent should exert minimal adverse effects on normal tissue and the capacity to kill tumor cells and/or inhibit tumor growth. However, most of classic chemotherapy agents are highly toxic, have a high economic cost and produce severe damage to healthy cells (Rajendran et al., 2014). Due to the increasing incidence of side effects related to cancer treatment as well as the development of resistance, classic chemotherapy is insufficient to provide a complete antitumor treatment and therefore, research has focused on the search for new drugs to provide an alternative therapy for supplementing or replacing conventional medicine. In this sense, natural products provide an incalculable source of chemical structures, and phytochemicals in particular play an important role in cancer therapies. In this context, many edible and medicinal plant-derived compounds have been associated with the chemoprevention and treatment of cancer. In recent decades, a high amount of research has been developed in order to discover natural compounds with potential antitumor activity and many of these new plant anticancer agents (e.g., paclitaxel, vinblastine, vincristine, topotecan, etoposide, etc.) have been used as successful cancer treatments. Among antitumor medications, 69% of all drugs approved between 1940 and 2002 are either natural products or developed based on knowledge obtained from natural products, a rate that is much higher than in other areas of drug developments (Sak, 2014).

Diet is closely related to the incidence and prevention of different cancer types and dietary behaviour has been identified as one of the most important modifiable lifestyle determinants of cancer risk. In fact, most authors agree that there are consistent epidemiological evidences to suggest that a diet rich in fruits and vegetables significantly reduces the risk of certain disorders, as cancer and cardiovascular diseases. Diet, together with a healthy lifestyle, can therefore reduce cancer incidence by 30-40%. In this sense, there is a demonstrated low cancer incidence in vegetarians (Calderón-Montaño et al., 2011; Cho et al., 2013; Guohua et al., 2011; Luo et al., 2011; Petrick et al., 2015; WHO, 2014). Many phytochemicals found in diet fruits and vegetables are regarded as cancer preventive agents. Flavonoids are the most important phytochemicals and their contribution to the prevention and treatment of cancer has been well studied (Sak, 2014). Flavonoids are natural products classified as safe, making them ideal candidates as cancer chemoprevention agents or associated agents in cancer co-therapy (Lim et al., 2007; Ninomiya et al., 2013; Szliszka et al., 2011).

Various *in vitro* and *in vivo* studies suggest that flavonoids have a diverse cytotoxic activity against several types of human cancer cells. This activity is mediated by different molecular targets and shows little or no effect on normal cells. These discoveries have stimulated the use of flavonoids as potential anticancer chemotherapeutics (Sak, 2014). However, the precise mechanisms responsible for this antitumor effect are not yet fully known and must be disclosed before any potential use in clinical practice. Therefore, there is still some work to carry out.

1.2. Kaempferol and Cancer

Kaempferol is an active ingredient isolated from different types of seeds and other plant derived products (honey, flowers, fruits, vegetables). Common dietary sources of kaempferol are beans (26 mg), bee olen (1.2 mg), broccoli (13 mg), cabbage (22 mg), capers (131 mg/100 g), cauliflower (0.5 mg), chia seeds (12 mg), chives (10 mg), cumin (0.2 mg), moringa leaves (6 mg), endive (10 mg), fennel (6 mg), garlic (0.3 mg), ginger (33 mg), kale (18 mg), leeks (1.6 mg), mizuna (6 mg), mustard (38 mg), onion (4 mg), parsley (1.5 mg), radish (21 mg), soybeans (1.3 mg), saffron (205 mg), spinach (6.3 mg), and black tea (1.4 mg) (Bhagwat et al., 2011; Rajendran et al., 2014). Kaempferol and its flavonoid derivatives (as glycosylated versions) exhibit a wide range of pharmacological activities, including antioxidant, antiinflammatory, antitumor, antimicrobial, cardioprotective, neuroprotective and antidiabetic activities (Calderón-Montaño et al., 2011).

Regular consumption of foods rich in this flavonol is positively associated with reduced risk of several types of disorders including cancer, based on diverse studies (Weng and Yen, 2012). A prospective study in the Dutch population has investigated the association between vegetables and fruits consumption and risk of esophageal squamous cell carcinoma (ESCC), esophageal adenocarcinoma (EAC), gastric cardia adenocarcinoma (GCA) and gastric non-cardia adenocarcinoma (GNCA). In this study, the role of individual groups of vegetables and fruits has been analyzed separately. It was observed significant inverse associations for raw vegetables and EAC risk (risk reduction per 25 g/day: 0.81, 95% confidence interval 0.68–0.98); and *Brassica* vegetables (known for their high content in kaempferol) and GCA risk (risk reduction per 25 g/day: 0.72, 95% confidence interval 0.54–0.95). This study concluded that consumption of specific groups of vegetables and fruits may protect against subtypes of esophageal and gastric cancer (Steevens et al., 2011).

Studies like this and a previous review by the American Institute for Cancer Research (AICR) about the benefits of increased cabbage consumption in advanced pancreatic cancer patients (American Institute for Cancer Research., 2007) led to a prospective, randomized, placebo-controlled clinical pilot study to evaluate, for the first time, the feasibility and clinical impact of a daily application of encapsulated, freeze-dried broccoli sprouts in pancreatic ductal adenocarcinoma patients in a palliative setting as a chemotherapy-supporting agent (Lozanovski et al., 2014).

The results found in the AICR study were not restricted to the pancreas, as two epidemiological, prospective nutrient studies demonstrated significant inhibition of metastases seeding in patients with prostate cancer (Kirsh et al., 2007; Richman et al., 2012). A significant lower risk of extra-prostatic manifestations for prostate cancer (stage III or IV tumors) correlated with an increase in the consumption of cruciferous vegetables, in particular broccoli, which is rich in sulforaphane and kaempferol (P = 0,02) (Kirsh et al., 2007). In another study, a double-blind, placebo-controlled randomized trial evaluated the effect of a polyphenol-rich whole food supplement on prostatic-specific antigen (PSA) progression in men with prostate cancer (Thomas et al., 2014). The ingredients of this interventional food supplement were pomegranate, turmeric, green tea and broccoli. These last two components are rich in kaempferol, as it is reported in the USDA Database for the Flavonoid Content of Selected Foods (Bhagwat et al., 2011). The PSA study found a significant short-term, favorable effect on the PSA percentage rise in men managed with active surveillance and watchful waiting following ingestion of this specific blend of concentrated foods.

A strong epidemiological evidence suggests that eating foods rich in the flavonol kaempferol could reduce the development of certain cancers, including lung cancer (Cui et al., 2008), gastric cancer (Garcia-Closas et al., 1999), pancreatic cancer (Nöthlings et al., 2007) and ovarian cancer (Gates et al., 2007). Moreover, these epidemiological studies have been proven by numerous *in vitro* experiments, and there are many examples in the literature that demonstrate their ability to inhibit the growth of different types of cancers, including glioma/glioblastoma (Sharma et al., 2007), breast adenocarcinoma (Diantini et al., 2012), leukemia (Ren et al., 2010), lung cancer (Leung et al., 2007), colorectal carcinoma (Li et al., 2009), pancreatic cancer (Zhang et al., 2008) and prostate cancer (Boam, 2015). It also has an antiproliferative activity and induces apoptosis in oral cavity cancer cells. This effect is enhanced because the oral cavity epithelium is highly permeable and can absorb directly flavonoids, achieving high levels of exposure to this phytochemical (Sak, 2014). But the most interesting feature of this flavonol is its high selectivity for tumor cells, showing little effect on non-cancerous cells (Matsuda et al., 2002).

In summary, epidemiological and nutritional intervention studies data suggest that a high intake of kaempferol-containing foods may reduce the risk of developing several types of cancers (gastric, pancreatic, lung, ovarian and prostatic cancer) and inflammatory processes associated with different diseases (as type 2 diabetes, see below). However, although many *in vitro* preclinical studies have shown that this flavonoid presents preventive and therapeutic properties against cancer, only a few studies have evaluated the antitumor potential of kaempferol in animal models (Yasukawa et al., 1990). The great potential of this flavonol has prompted scientists to investigate the molecular mechanisms involved in its antitumor actions, in order to assess its value for cancer treatment. Various *in vitro* and *in vivo* studies have provided broad evidence for kaempferol as an agent able to prevent carcinogenesis and to inhibit tumorigenesis through different molecular mechanisms (Chen and Charlie, 2013).

1.3. Molecular Targets of Kaempferol and Their Relationship with the Hallmarks of Cancer

It is widely accepted that the formation of a malignant tumor requires neoplastic cells to acquire a range of skills that are known as hallmarks of cancer (Hanahan and Weinberg, 2011) (Figure 1). Kaempferol makes a direct impact on some of these hallmarks, being able to induce apoptosis and thereby releasing tumor cells resistance to it (Huang et al., 2010; Kang et al., 2009; Marfe et al., 2009; Sharma et al., 2007). Kaempferol also shows an *in vitro* inhibitory activity on angiogenesis (generation of new blood vessels) (Kim et al., 2006; Schindler and Mentlein, 2006). Metastasis from primary tumors is responsible for 90% of all cancer deaths. Metastasis is the ability of tumor cells to invade the adjacent tissue and to travel to other organs, where they are able to generate secondary tumors. This is one of the major processes involved in carcinogenesis and another of the hallmarks of cancer. In this regard, kaempferol can inhibit this process both *in vitro* and *in vivo* (Labbé et al., 2009). A common factor to all these hallmarks is genomic instability in tumor cells, a process that generates genomic diversity in tumors, suggesting that cancer is caused by alterations in DNA as submitted by the somatic mutation theory of cancer (Hanahan and Weinberg, 2011; Vogelstein and Kinzler, 2004). Many studies suggest that low kaempferol levels may protect the DNA from damage induced by different carcinogens (Cemeli et al., 2004).

1.4. Effects on the Cell Cycle

Kaempferol is an inhibitor of cancer cell proliferation and an apoptosis inducer in various tumors, including lung, pancreas and ovary cancer. Its *in vivo* effect has been studied using subcutaneous xenografts in mice. Kaempferol *in vivo* antitumor effects have been evaluated in BALB/c nude mice inoculated with human osteosarcoma cells and the results showed a clear inhibition of tumor growth (Huang et al., 2010). Both induction of apoptosis and cell cycle arrest were observed in *in vitro* models, and these data were corroborated in *in vivo* studies, demonstrating a suppression of tumor growth (compared with control group), a decrease in multiplication related markers and an increase in apoptosis markers (Dang et al., 2015).

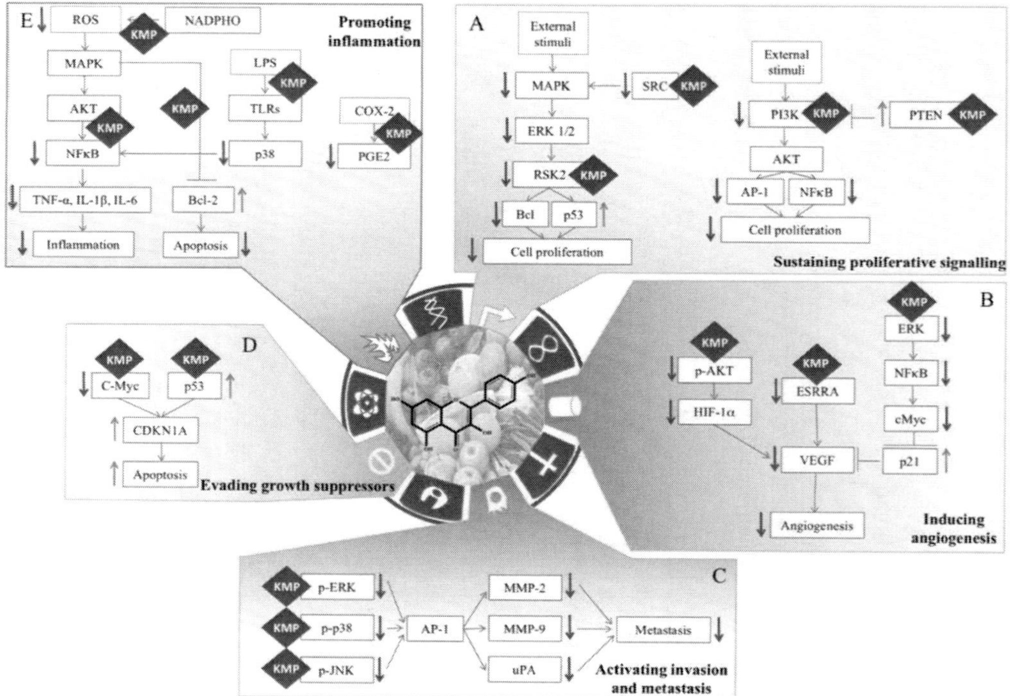

Figure 1. Cellular targets of kaempferol affecting diverse transduction pathways of interest in cancer: sustained proliferating signalling, apoptosis induction, metastasis activation, growth suppressors' evasion and inflammation promotion (modified from Hanahan and Weinberg, 2011).

The cell cycle is tightly regulated by cyclins and cyclin-dependent kinases (CDKs), which form the mitosis-promoting factor (MPF) that mediates the passage of the G_2 to the M phase. MPF is a target of great interest in cancer treatment to arrest the cell cycle and thus cell growth (Chen and Charlie, 2013). Tumor cells are characterized by a *c-myc* oncogene overexpression leading to an uncontrolled proliferation by regulating the AP4 expression, as well as decreased CDKN1A levels, whose function is to produce cell cycle arrest (Jung et al., 2008; Luo et al., 2010a). When kaempferol was added to ovarian cancer cells *in vitro* cultures, *c-myc* mRNA levels decreased and CDKN1A levels increased, due to the elimination of the antagonistic effect of c-Myc in CDKN1A, leading to cell cycle arrest (Figure 1D). Increased CDKN1A levels can be performed by another complementary pathway where p53 is involved, a well-known tumor suppressor protein that is able to activate CDKN1A when detecting DNA damage. Kaempferol stimulates p53 activation *in vitro,* and therefore apoptosis of breast cancer and cervical cancer HeLa cells (Ito et al., 2004; Kang et al., 2009; Xu et al., 2008).

1.5. Effects on Signal Transduction

Cells are able to respond to a wide range of external stimuli by using transduction pathways leading to changes in gene expression. One of these transduction pathways is the MAPK/ERK pathway that performs an integral role by promoting the regulation of cell growth, but it is also required in cell death (Kim et al., 2008). The apoptosis-inducing activities of kaempferol are believed to be due to its effects on the MAPK pathway: (a) it is able to bind directly to the active pocket of RSK2 (an apoptosis suppressor protein whose ectopic expression is related to neoplastic transformation) inhibiting its action (Cho et al., 2009; Yoo et al., 2015), resulting in decreased Bcl levels and increased levels of tumor suppressor proteins as BAD and p53 (Figure 1A) (Luo et al., 2011); (b) kaempferol is able to disrupt Src kinase activity acting as a competitive inhibitor and preventing its activation by ATP, blocking its skin cancer promoting activity. The Src protein is able to activate the MAPK/ERK pathway resulting in an increased cell proliferation (Luo et al., 2011). Once kaempferol has neutralized both apoptosis suppressors (RSK2 and Src), apoptosis can proceed at a faster pace.

Another transduction pathway involved in tumor development is PI3K/AKT. PI3K overactivation leads to an accumulation of AKT that regulates various transcription factors, as AP-1 and NF-κB, which lead to the carcinogenesis (Li et al., 1997; Nomura et al., 2003). In this case, kaempferol also performs a competitive effect on the binding of ATP to PI3K, as demonstrated in mouse epidermal cells, therefore inhibiting downstream AKT activity on its target transcription factors AP-1 and NF-κB (Figure 1A) (Lee et al., 2010a). As in the previous case, the apoptotic process starts when their repressors are inhibited (Ruiz et al., 2006). Other studies have demonstrated that kaempferol induces apoptosis in human bladder cancer cells by increasing the activity of the tumor suppressor PTEN and by decreasing the phosphorylation of AKT. In this schema, apoptosis induction by kaempferol was significantly reduced in PTEN-knockdown cell lines (Xie et al., 2013). Finally, kaempferol downregulates PI3K/AKT and ERK-1/2 pathways in human colon cancer cells, inhibiting IGF-IR and ErbB3 signalling (Figure 1A) (Lee et al., 2014).

1.6. Effects on Apoptosis

Kaempferol exerts a direct effect on the apoptosis extrinsic pathway, which is based on the presence of death receptors on the cell surface able to recognize death inducing substances. These death receptors include tumor necrosis factor alpha (TNF-α), FAS and TRAIL (Thorburn, 2004). TRAIL receptor is of particular interest, since it induces apoptosis in human colon cancer cells and a deficiency in its expression on cell surface explains tumor resistance to apoptosis (Jin et al., 2004). Kaempferol is able to up-regulate TRAIL receptors by decreasing the resistance of tumor cells to apoptosis and sensitizing cells towards TRAIL-dependent apoptosis (Yoshida et al., 2008).

1.7. Effects on Epigenetic Markers

In recent years, an increasing number of nutritional components that have an inherent epigenetic activity have been identified. These micronutrients are able to influence gene expression by carrying out an inheritable modification of DNA (or DNA-associated proteins) without changing the DNA sequence. One of the best known epigenetic mechanisms is the posttranslational modification of histone proteins by histone deacetylases (HDACs). Certain tumors exhibit an overexpression of HDACs, contributing to the shut-down of regulatory genes. Therefore, discovery of new HDACs inhibitors (HDACIs) is of great interest as potential anticancer drugs (Berger et al., 2013). In this sense, kaempferol possesses epigenetic activity as HDACI by joining its HDACs binding pocket. Kaempferol induces therefore hiperacetylation of histone complexes and therefore causes cell viability and proliferation reduction in hepatoma and colon cancer cell lines (Berger et al., 2013).

1.8. Effects on Angiogenesis

Tumor cells require oxygen and nutrients, and in order to achieve these nutrients, they form new blood vessels around by a process called angiogenesis. The main mediator of this process is vascular endothelial growth factor (VEGF) (Ferrara, 2004).

Kaempferol has demonstrated antiangiogenic activities (Kim and Choi, 2013). For example, it can reduce tumor proliferation by significantly reducing VEGF expression in ovarian cancer cells, both at its mRNA gene expression and protein levels. This angiogenesis inhibition is achieved by downregulating hypoxia-inducible factor 1 (HIF-1), which is a key regulator of oxygen homeostasis and possesses a role in several important aspects of carcinogenesis, including angiogenesis, invasion, metastasis and resistance to apoptosis (Semenza, 2006, 2007). There is a direct relation between HIF-1 overexpression and increased tumor aggressiveness and therefore patient mortality. In fact, some authors have detected HIF-1 overexpression in various tumors including lung, breast, colon and prostate, which are the most common cancers in Western countries. These data suggest that HIF-1 activation is a key event in carcinogenesis and therefore a potential therapeutic target (López-Lázaro, 2006; Zhong et al., 1999). Different studies in hepatoma and ovarian cancer showed that kaempferol is able to inhibit HIF-1 transcription factor at a low micromolar range (Luo et al., 2009; Mylonis et al., 2010)

Kaempferol can also inhibit angiogenesis in a dose-dependent manner by an independent route by downregulating ESRRA, which is a VEGF-independent HIF-1 regulator (Luo et al., 2009). Therefore, this compound is able to inhibit angiogenesis and VEGF expression through both HIF-1-dependent (AKT/HIF-1) and HIF-1-independent (ESRRA) pathways (Figure 1B) (Luo et al., 2009). Recently it has been described another pathway modulated by this flavonol, eliminating extracellular signal-regulated kinase (ERK)-NFκB-cMyc-p21-VEGF pathway in ovarian cancer cells (Luo et al., 2012a).

1.9. Effects on Metastasis

In order to spread to other tissues, tumor cells must first degrade the surrounding extracellular matrix (ECM) to reach the blood vessels and then propagate throughout the body. Tumor cells use a variety of enzymes for this purpose, as the matrix metalloproteinases (MMPs), which are often associated to poor clinical prognoses in cancer patients (Guan, 2015).

Kaempferol can inhibit cell metastasis through ERK-p38, JNK and AP-1 signalling pathway in human osteosarcoma cells. This flavonol is able to reduce protein phosphorylations at ERK, p38 and JNK, therefore decreasing the DNA binding activity of AP-1, and causing reduced expression of MMP-2, MMP-9 and uPA (urokinase-type plasminogen activator), therefore reducing the metastatic potential (Figure 1C) (Chen et al., 2013; Li et al., 2014; Lin et al., 2013).

2. Antiinflammatory Activity of Kaempferol on Diverse Diseases

Reactive oxygen species (ROS) play an important role in many inflammatory diseases and in carcinogenesis, as tumor cells commonly have higher ROS levels, inducing malignant transformation phenotypes that can be reversed when ROS cellular levels are reduced (Arnold et al., 2001; López-Lázaro, 2006). Therefore, antioxidants that can prevent accumulation and/or reduce ROS cellular levels perform a protective role in inflammatory disorders and cancer development, being able to reverse their phenotype and normalizing their growth rate (Arnold et al., 2001). In this sense, kaempferol has a widely demonstrated antioxidant and anti-inflammatory activities in many diseases. Also, long term inflammation has been described as a cause of cancer development in some situations, as in *Helicobacter pylori* and hepatitis B virus infections and gastric and hepatic cancers, respectively (Grivennikov et al., 2010). Under these inflammatory conditions, ROS trigger activation of MAPKs (as ERK, JNK, p38) and subsequently of the pro-inflammatory NF-κB transcription factor, giving rise to a diverse production of pro-inflammatory cytokines (TNF-α, IL-1β, IL-6, etc.). Kampferol has been studied as an antioxidant flavonoid and anti-inflammatory nutraceutical in diverse diseases, with interesting results (Chen and Chen, 2013; Rajendran et al., 2014; Devi et al., 2015).

A nutritional intervention trial evaluated the effect of cruciferous vegetables on systemic inflammation biomarkers in healthy adults. IL-6 concentrations were significantly lower on day 14 of two different cruciferous diets than with the basal diet (219% lower (9 5% CI: 230%, 20.1%) and 220% lower (95% CI: 231%, -0.7%), respectively). IL-8 concentrations were higher after one cruciferous diet (+16%; 95% CI: 4.2%, 35.2%) than after the basal diet and there were no effects of diet on CRP or TNF-α (Navarro et al., 2014).

In vitro, 25 μM kaempferol has been able to diminish inflammation parameters and to enhance cell survival in HT22 neuron cell lines challenged with glutamate. Glutamate is a neurotransmitter which in high doses induces ROS and neurotoxicity. This increased cell mortality is caused by higher levels of apoptosis due to lower Bcl-2 anti-apoptotic protein in glutamate treated neurons. The kaempferol neuroprotective effect is based on anti-apoptosis activity, as this flavonol recovers the cytosol concentration of Bcl-2 even under high glutamate concentrations. Kaempferol could therefore be used for preventing neuronal damage in inflammatory conditions as Alzheimer or Parkinson's diseases (Yang et al., 2014). Interestingly, the treatment of glioblastoma cells with kaempferol caused an increase in ROS due to lower levels of superoxide dismutase and thioredoxin; but also an increase in apoptosis by decreasing expression of Bcl-2. Also, in this *in vitro* assay, kaempferol showed a synergistic effect with the antitumor doxorubicin, by increasing more the ROS intracellular pool (Sharma et al., 2007).

However, the pro-apoptotic effect of doxorubicin *in vivo* on normal cells (cardiomiocytes) was prevented in a rat model by intraperitoneal kaempferol administered before the antitumor treatment. In this case, the preventive effect was due to an inhibition of p53 activation, therefore preventing apoptosis due to Bax expression inhibition (Xiao et al., 2012). Kaempferol, at 25 μg/ml also showed a protective anti-inflammatory activity on cardiac fibroblasts after a challenge with LPS, suppressing the production of pro-inflammatory cytokines as TNF-α, IL-1β, IL-6 and IL-18 (Figure 1E). This protective effect against cardiovascular inflammation is due to the inhibition of Akt and NFκB cascades and may be of interest for *in vivo* prevention of atherosclerosis and other cardiovascular diseases (Figure 1E) (Tang et al., 2015). In this sense, an *in vivo* assay for atherosclerosis in rabbits getting a high cholesterol diet for ten weeks showed that oral kaempferol (150 mg/kg body weight) caused a marked protection in the formation of aortic atherosclerosis, as well as a reduction in biomarkers for this disease (malondialdehyde, LDL, TNF-α, IL-1β) (Kong et al., 2013).

Several case-control and cohort studies have evaluated the relationship between consumption of kaempferol-rich foods and the inflammatory status in healthy adults and diabetic patients. A nutritional intervention clinical study in male smokers evaluated the effects of broccoli intake (250 g/day) during 10 days on dietary and inflammation markers. This study found that broccoli intake significantly increased folate (+17%) and lutein (+39%) plasma levels, while no significant effect was observed for TNF-α, IL-6 nor adiponectin. On the other hand, plasma C-reactive protein (CRP) decreased in 48% (post-hoc analysis, $p < 0.05$) following this broccoli diet (Riso et al., 2014). Another nutritional intervention trial evaluated the influence of two diets on inflammatory biomarkers among type 2 diabetic patients: non-soya legume-based therapeutic lifestyle change diet (TLC) versus isoenergetic legume-free TLC diet. The legume based diet had lentils, chickpeas, peas and beans, the last ones being highly reach in kaempferol. Compared with the legume-free TLC diet, the non-soya legume-based TLC diet significantly decreased CRP, IL-6 and TNF-α in the overweight diabetic patients (Hosseinpour-Niazi et al., 2015). Also on type 2 diabetes patients, it has been evaluated the efficacy of consuming a caper extract (high in kaempferol), using a randomized double-blind placebo-controlled clinical trial. Results showed significant decrease in fasting blood glucose levels ($p = 0.037$) and glycosylated hemoglobin ($p = 0.043$) in caper treated patients compared to control group at the end of the study. Triglycerides level also decreased significantly ($p = 0.29$) in caper treated group compared to baseline (Fallah et al., 2013).

Oral administration of kaempferol (100 mg/kg of body weight) to streptozotocin-treated rats (a common animal model for diabetes) was able to reduce plasma glucose levels (-65%) and plasma and tissues lipid peroxidative markers, and at the same time, to increase plasma insulin levels (+110%). In these rats, kaempferol caused also an increase in enzymatic and non-enzymatic antioxidants (glutathione peroxidase, superoxide dismutase, catalase, glutathione, vitamins C and E), improving the ROS status. Kaempferol can therefore prevent oxidative damage in pancreatic beta cells and improves glucose uptake by adipocytes (Al-Numair et al., 2015). In this sense, four weeks administration of broccoli sprouts (a kaempferol-rich food) to type 2 diabetic patients (10 g/day) resulted in a significant increase in total antioxidant capacity (+ 10%), and lower LDL cholesterol (-5%) and glucose (-22%) levels (Bahadoran et al., 2011).

The cellular targets affected by kaempferol in these diabetic animal models have been studied in the streptozotocin rat model, including a high fat diet in order to cause also hepatic inflammation. Oral administration of 150 mg/kg body weight in these rats caused a restoration to normal control levels in blood lipids and insulin parameters. Kaempferol inhibited phosphorylation of IRS-1 (insulin receptor substrate 1), IKKα (IkB kinase α) and IKKβ (IkB kinase β), three proteins belonging to an important inflammation pathway, where streptozotocin acts by triggering their phosphorylations (activations). Also, a reduction in cytoplasm and nucleus active NF-κB (free from IκBα or IκBβ) was observed in these hepatocytes, also demonstrating a lower inflammatory status (Figure 1E). This was confirmed by decreased levels of pro-inflammatory elicitors TNF-α and IL-6, which under diabetic conditions promote IRS-1 phosphorylation again (Kim et al., 2016; Luo et al., 2015). *In vivo,* kaempferol (4 mg/kg body weight) was able to prevent age-related NF-κB activation in rat kidney, in response to advance glycation end-products (AGE); an inhibition caused by blocking NADPH oxidation generation of ROS. NADPH oxidase is the response enzyme to AGE in these cells, causing production of H_2O_2 and superoxide (Figure 1E) (Kim et al., 2010).

Liver protection has been described for kaempferol by using a completely different experimental approach, using this flavonol for prevention of hepatotoxicity caused by compounds as ethanol or the antibiotic rifampicin. Liver metabolism of these compounds and other drugs involves the cytochrome enzyme CYP2E1, and this metabolism causes an increase in ROS parameters as malondialdehyde and transaminases; with reduced levels of antioxidant enzymes. *In vivo* treatments with this flavonol in mice caused a recovery in hepatotoxicity parameters, with a direct inhibition on CYP2E1 (Shih et al., 2013; Wang et al., 2015).

Kaempferol has shown activity also in animal models for another inflammatory condition, asthma. In a mouse model (intraperitoneal sensitization with ovalbumin/alum and challenged with 0.4 ml/min aerosol containing 2% ovalbumin), this flavonol was able to reduce the number of inflammatory cells in bronchoalveolar lavage fluid (BALF), and also inhibited Akt phosphorylation (almost to normal control levels), giving rise to a reduction in pro-inflammatory TNF-α (-20%), IL-4 (45%), IL-5 (-25%) and IL-13 (-55%) cytokine levels. Also, it inhibited eosinophilia, therefore reducing IgE levels by 33% in asthmatic mice. All these molecular and cellular parameters are elevated in asthma and contribute to their symptoms (Chung et al., 2015). This protective effect has also been demonstrated in *in vitro* studies with eosinophils treated with kaempferol (Oh et al., 2013). *In vivo*, this flavonol, at oral doses of 20 mg/kg body weight, is able to block STAT3 signaling pathway activation in response to ovalbumin antigen (but also in response to LPS) (Gong et al., 2013).

Kaempferol at an oral dosis of 100 mg/kg body weight was able to ameliorate acute lung inflammation in a mice model (through intranasal instillation of lipopolysaccharide, LPS). Specifically, this flavonol reduced BALF levels for pro-inflammatory IL-1β, IL-6 and TNF-α, increasing superoxide dismutase levels. These effects were carried out through inhibition TLRs (toll-like receptors) activation by LPS, with subsequent inhibition of diverse MAPKs, specifically JNK and p38, causing a diminished degradation of IκBβ and therefore lower NF-κB levels (Figure 1) (Chen et al., 2012). One of the effects of IL-1β in the lung epithelium is to increase mucin expression via ERK and P38 MAPKs pathways, an important contributor to symptoms. *In vitro*, kaempferol (at 80 μM) is able to block mucin expression in NCI-H292 cells (human pulmonary mucoepidermoid carcinoma) by inhibiting these MAPKs pathways (Kwon et al., 2009). *In vivo*, this flavonol is able to block TLR4 activation by LPS in mice airways (Gong et al., 2013). NF-κB inhibition by kaempferol has also been described *in vitro* for Jurkat leukemia cells after 72 h treatment with 10 μM flavonol (Kadioglu et al., 2015).

In a mouse model for ulcerative colitis (using oral dodecyl sodium sulfate, DSS), oral administration of 0.3% kaempferol was able to enhance the intestinal health. DSS causes an increase in intestinal permeability and mucosal damages due to an increase in pro-inflammatory NO and PGE_2 levels. In the kaempferol treated animals, intestinal mucosa PGE_2 and NO levels diminished, results that were confirmed by detecting lower expression of their biosynthetic enzymes COX-2 and iNOS (Figure 1E) (Park et al., 2012). The same effect on COX-2 and iNOS has been observed by other authors in a rat model for air pouch inflammation in the back, with oral kaempferol (100mg/kg body weight) (Yunnus et al., 2010), in a mouse model for UVB inflammation (Lee et al., 2010b), and in an *in vitro* inflammation model using macrophages that were challenged with LPS (Kim et al., 2015), proving these two enzymes as kaempferol targets. The same protective effect against LPS-mediated inflammation via macrophages has been described for kaempferol in a mouse model for mastitis, with reduced activation of NF-κB and reduced levels of secreted IL-1β, IL-6 and TNF-α (Figure 1E) (Cao et al., 2014). In another allergic inflammation *in vitro* model using IgE-stimulated Caco-2 colon cells, kaempferol was able to inhibit ERK activation and IL-4 and TNF-α production (Lee et al., 2010c).

In a rat model for colorectal cancer, oral administration of kaempferol (200 mg/kg body weight) caused a recovery in some antioxidant enzymes as glutathione peroxidase and superoxide dismutase, a chemoprevention activity against this cancer (Nirmala and Ramanathan, 2011). A well known apoptosis inducer on colon cancer cells is butyrate (as well as other short-chain fatty acids), which possesses a strong HDACs activity. Interestingly, kaempferol has also been described as HDAC inhibitor in *in vitro* assays against HCT-116 colon cancer lines at 5 μM concentration, which opens its use as epigenetic modulator on this and other cancers (Berger et al., 2013).

3. Importance of Flavonoids in Co-Therapy

Kaempferol has also a great potential in combination with antitumor drugs for enhancing their therapeutic effects, a practice which is known as co-therapy. The combination of kaempferol with classical chemotherapeutic agents results in greater cytotoxic effects than those achieved by each of them separately (Luo et al., 2010a). This flavonol is able to sensitize tumor cells in the presence of the cytotoxic effects of cisplatin (Luo et al., 2010b), 5-fluorouracil (Zhang et al., 2008), doxorubicin (Sharma et al., 2007) and cytarabine (Nadova et al., 2007).

Flavonoids as kaempferol are recognized by the body as foreign substances and cells possess pumps that secrete these xenobiotic compounds out of their cytoplasm (Schinkel and Jonker, 2003). One of these pumps is the breast cancer resistant protein (ABCG2) and flavonoids such as quercetin or kaempferol have affinity for it at different degrees. Another highly promising role for kaempferol in co-therapy that differs drastically from other flavonoids (as isoflavones) is its ability to significantly decrease Pgp (P-glycoprotein 1) expression in a dose-dependent manner in certain human cervical carcinoma cells. Kaempferol carries out this by increasing the intracellular accumulation of the antitumor vinblastine, therefore contributing to a reduction in its chemotherapy doses. This effect is not exerted in cells lacking Pgp, suggesting a relationship between kaempferol and the expression of this protein (Limtrakul et al., 2005). Pgp carries out the efflux of chemotherapeutic drugs from cells and it is an important mechanism in multidrug resistance (MDR) in human tumors (Limtrakul et al., 2005). Simultaneously administration of both quercetin and kaempferol blocks quercetin efflux to the extracellular medium because kaempferol exhibits a greater affinity for the pump, allowing quercetin to remain within the cell and performing its antitumor effect (Guohua et al., 2011). The importance of co-therapy has also been shown on the joint implementation of kaempferol and doxorubicin in glioma cells, where the flavonol enhanced the toxic effect of the chemotherapeutic agent doxorubicin and decreased its cellular efflux (Sharma et al., 2007). In rats, oral co-administration of tamoxifen and kaempferol showed an increase in tamoxifen bioavailability, because this drug is a substrate for Pgp and CYP3A (microsomal cytochrome P_{450}), and both of them are specifically inhibited by kaempferol (Piao et al., 2008). Kaempferol is also able to decrease the levels of ABCC6 mRNA, another ATB-binding cassette transporter, potentiating the antitumor effects of cisplatin (Luo et al., 2010a). Therefore, kaempferol could be combined with other drugs which have greater affinity for ABCC6.

3.1. Limitations of Kaempferol as a Chemotherapeutic Agent

Many researchers have focused on understanding the *in vitro* effects of dietary flavonoids including kaempferol. But the question which remains is whether kaempferol is as effective in cancer patients as it is in the *in vitro* and *in vivo* models. Epidemiology data support this idea as a general concept, as it is known that a low fruits and vegetables intake is associated with an increased risk of cancer (Chen and Charlie, 2013). However, other authors do not see a clear relationship based on other epidemiology data (Gates et al., 2007; Wang et al., 2009). This incongruity between the impressive *in vitro* and *in vivo* studies of kaempferol and certain epidemiological evidences can be explained by the low bioavailability of all flavonoids, because these nutraceuticals are subjected to first-pass metabolism through the intestinal wall (an important barrier for highly hydroxylated compounds as these ones) and the enterohepatic circulation (which generates less active derivatives) (Barve et al., 2009).

Several studies have focused on measuring peak plasma concentrations after an oral intake of defined quantities of kaempferol, and its maximum plasma concentration has been accomplished after ingesting a bowl of thick endive soup containing 9 mg of kaempferol. In this experiment, the recorded kaempferol blood peak was 150 nM (DuPont et al., 2004), far away from the 5 μM necessary to achieve an inhibition on tumor cell proliferation as in *in vitro* experiments (Berger et al., 2013). Therefore, to exploit kaempferol benefits, this compound needs to be administered in higher concentrations than it appears in food, without exceeding the toxicity limit. One of the most promising techniques for improving flavonoids bioavailability is by using nanoparticles. These increase the nutraceutical permeability without affecting its chemical structure (Luo et al., 2012b). Nanoparticles have a great potential in biomedicine as nutraceuticals and drugs delivery systems because of their small size and large surface to volume ratio (Srinivas Raghavan et al., 2015).

Despite the proven effectiveness of kaempferol in cell lines and animal models, there are very few clinical studies to date. Therefore, further *in vivo* studies are necessary in order to consider kaempferol as a drug or co-drug for the treatment of different types of carcinomas (Rajendran et al., 2014) and the use of nanoparticles can be of great help in this.

References

Al-Numair KS, Chandramohan G, Veeramani C and Alsaif MA (2015) Ameliorative effect of kaempferol, a flavonoid, on oxidative stress in streptozotocin-induced diabetic rats. *Redox Report : Communications in Free Radical Research* 20:198–209.

American Institute for Cancer Research (2007) *Food, Nutrition, Physical Activity, and the Prevention of Cancer: a Global Perspective.*

Bahadoran Z, Mirmiran P, Hosseinpanah F, Hedayati M, Hosseinpour-Niazi S and Azizi F (2011) Broccoli sprouts reduce oxidative stress in type 2 diabetes: a randomized double-blind clinical trial. *European Journal of Clinical Nutrition* 65:972–977.

Kong A-N (2009) Metabolism, oral bioavailability and pharmacokinetics of chemopreventive kaempferol in rats. *Biopharmaceutics & Drug Disposition* 30:356–365.

Berger A, Venturelli S, Kallnischkies M, Böcker A, Busch C, Weiland T … Bitzer M (2013) Kaempferol, a new nutrition-derived pan-inhibitor of human histone deacetylases. *The Journal of Nutritional Biochemistry* 24:977–985.

Bhagwat S, Haytowitz DB and Holden JM (2011) Database for the Flavonoid Content of Selected Foods Release 3 Prepared by USDA Database for the Flavonoid Content of Selected Foods Release 3. *U.S. Department of Argiculture* 1–156.

Boam T (2015) Anti-androgenic effects of flavonols in prostate cancer. *Ecancermedicalscience* 9:585.

Calderón-Montaño JM, Burgos-Morón E, Pérez-Guerrero C and López-Lázaro M. (2011) A review on the dietary flavonoid kaempferol. *Mini Reviews in Medicinal Chemistry* 11:298–344.

Cao R, Fu K, Lv X, Li W and Zhang N (2014) Protective effects of kaempferol on lipopolysaccharide-induced mastitis in mice. *Inflammation* 37:1453–145.

Cemeli E, Schmid TE and Anderson D (2004) Modulation by flavonoids of DNA damage induced by estrogen-like compounds. *Environmental and Molecular Mutagenesis* 44:420–426.

Chen AY and Charlie Y (2013) A review of the dietary flavonoid , kaempferol on human health and cancer chemoprevention. *Food Chemistry* 138:2099–2107.

Chen AY and Chen YC (2013) A review of the dietary flavonoid, kaempferol on human health and cancer chemoprevention. *Food Chemistry* 138:2099–2107.

Chen HJ, Lin CM, Lee CY, Shih NC, Peng SF, Tsuzuki M … Yang JS (2013) Kaempferol suppresses cell metastasis via inhibition of the ERK-p38-JNK and AP-1 signaling pathways in U-2 OS human osteosarcoma cells. *Oncology Reports* 30:925–932.

Chen X, Yang X, Liu T, Guan M, Feng X, Dong W … Sun Z (2012) Kaempferol regulates MAPKs and NF-κB signaling pathways to attenuate LPS-induced acute lung injury in mice. *International Immunopharmacology* 14:209–216.

Cho HJ, Han J and Park Y (2013) Kaempferol Induces Cell Cycle Arrest in HT-29 Human Colon Cancer Cells. *Journal of Cancer Prevention* 18:1–7.

Cho Y-Y, Yao K, Pugliese A, Malakhova ML, Bode AM and Dong Z (2009) A regulatory mechanism for RSK2 NH(2)-terminal kinase activity. *Cancer Research* 69:4398–4406.

Chung MJ, Pandey RP, Choi JW, Sohng JK, Choi DJ and Park YIl (2015). Inhibitory effects of kaempferol-3-O-rhamnoside on ovalbumin-induced lung inflammation in a mouse model of allergic asthma. *International Immunopharmacology* 25:302–310.

Cui Y, Morgenstern H, Greenland S, Tashkin DP, Mao JT, Cai L … Zhang, Z-F (2008) Dietary flavonoid intake and lung cancer--a population-based case-control study. *Cancer* 112:2241–2248.

Dang Q, Song W, Xu D, Ma Y, Li F, Zeng J … Li L (2015) Kaempferol suppresses bladder cancer tumor growth by inhibiting cell proliferation and inducing apoptosis. *Molecular Carcinogenesis* 54:831-840.

Devi KP, Malar DS, Nabavi SF, Sureda A, Xiao J, Nabavi SM and Daglia M (2015) Kaempferol and inflammation: From chemistry to medicine. *Pharmacological Research* 99:1–10.

Diantini A, Subarnas A, Lestari K, Halimah E, Susilawati Y, Supriyatna JE … Abdulah R (2012) Kaempferol-3-O-rhamnoside isolated from the leaves of Schima wallichii Korth. inhibits MCF-7 breast cancer cell proliferation through activation of the caspase cascade pathway. *Oncology Letters* 3:1069–1072.

DuPont MS, Day AJ, Bennett RN, Mellon FA and Kroon PA (2004) Absorption of kaempferol from endive, a source of kaempferol-3-glucuronide, in humans. *European Journal of Clinical Nutrition* 58:947–954.

Fallah H, Hasani-rnjbar S, Nayebi N, Heshmat R, Khaliqi F, Ahvazi M … Kianbakht S (2013) Capparis spinosa L. (Caper) fruit extract in treatment of type 2 diabetic patients : A randomized double-blind placebo-controlled clinical trial. *Complementary Therapies in Medicine* 21:447–452.

Ferrara N (2004). Vascular endothelial growth factor as a target for anticancer therapy. *The Oncologist* 9 Suppl 1:2–10.

Garcia-Closas R, Gonzalez CA, Agudo A and Riboli E (1999) Intake of specific carotenoids and flavonoids and the risk of gastric cancer in Spain. *Cancer Causes & Control :CCC* 10:71–75.

Gates MA, Tworoger SS, Hecht JL, De Vivo I, Rosner B and Hankinson SE (2007) A prospective study of dietary flavonoid intake and incidence of epithelial ovarian cancer. *International Journal of Cancer. Journal International Du Cancer* 121:2225–2232.

Gong J, Shin D, Han S, Park S, Kang M, Kim J and Kang Y (2013) Blockade of Airway Inflammation by Kaempferol via Disturbing Tyk-STAT Signaling in Airway Epithelial Cells and in Asthmatic Mice. *Evidence-Based Complementary & Alternative Medicine* 2013:250725.

Grivennikov SI, Greten FR and Karin M (2010) Immunity, inflammation, and cancer. *Cell* 140:883–899.

Guan X (2015) Cancer metastases: challenges and opportunities. *Acta Pharmaceutica Sinica B* 5:402–418.

Guohua A, Gallegos J and Morris ME (2011) The bioflavonoid kaempferol is an Abcg2 substrate and inhibits Abcg2-mediated quercetin efflux. *Drug Metabolism and Disposition: The Biological Fate of Chemicals* 39:426–432.

Hanahan D and Weinberg RA (2011) Hallmarks of cancer: the next generation. *Cell* 144:646–674.

Hosseinpour-Niazi S, Mirmiran P, Fallah-Ghohroudi A and Azizi F (2015) Non-soya legume-based therapeutic lifestyle change diet reduces inflammatory status in diabetic patients: a randomised cross-over clinical trial. *The British Journal of Nutrition* 114:213–219.

Huang W-W, Chiu Y-J, Fan M-J, Lu H-F, Yeh H-F, Li K-H … Yang J-S (2010) Kaempferol induced apoptosis via endoplasmic reticulum stress and mitochondria-dependent pathway in human osteosarcoma U-2 OS cells. *Molecular Nutrition & Food Research* 54:1585–1595.

Ito K, Nakazato T, Yamato K, Miyakawa Y, Yamada T, Hozumi N … Kizaki M (2004) Induction of apoptosis in leukemic cells by homovanillic acid derivative, capsaicin, through oxidative stress: implication of phosphorylation of p53 at Ser-15 residue by reactive oxygen species. *Cancer Research* 64:1071–1078.

Jin Z, McDonald ER, Dicker DT and El-Deiry WS (2004) Deficient tumor necrosis factor-related apoptosis-inducing ligand (TRAIL) death receptor transport to the cell surface in human colon cancer cells selected for resistance to TRAIL-induced apoptosis. *The Journal of Biological Chemistry* 279:35829–35839.

Jung P, Menssen A, Mayr D and Hermeking H (2008) AP4 encodes a c-MYC-inducible repressor of p21. *Proceedings of the National Academy of Sciences of the United States of America* 105:15046–15051.

Kadioglu O, Nass J, Saeed MEM, Schuler B and Efferth T (2015) Kaempferol Is an Anti-Inflammatory Compound with Activity towards NF-κB Pathway Proteins. *Anticancer Research* 35:2645–2650.

Kang G-Y, Lee E-R, Kim J-H, Jung JW, Lim J, Kim SK … Kim KP (2009) Downregulation of PLK-1 expression in kaempferol-induced apoptosis of MCF-7 cells. *European Journal of Pharmacology* 611:17–21.

Kim B-W, Lee E-R, Min H-M, Jeong H-S, Ahn J-Y, Kim J-H … Cho S-G (2008) Sustained ERK activation is involved in the kaempferol-induced apoptosis of breast cancer cells and is more evident under 3-D culture condition. *Cancer Biology & Therapy* 7:1080–1089.

Kim JM, Lee EK, Kim DH, Yu BP and Chung HY (2010) Kaempferol modulates pro-inflammatory NF- κ B activation by suppressing advanced glycation endproducts-induced NADPH oxidase. *Age* 32:197–208.

Kim J-D, Liu L, Guo W and Meydani M (2006) Chemical structure of flavonols in relation to modulation of angiogenesis and immune-endothelial cell adhesion. *The Journal of Nutritional Biochemistry* 17:165–176.

Kim O-K, Jun W and Lee J (2016) Effect of Cudrania tricuspidata and Kaempferol in Endoplasmic Reticulum Stress-Induced Inflammation and Hepatic Insulin Resistance in HepG2 Cells. *Nutrients* 8:60.

Kim SH, Park JG, Lee J, Yang WS, Park GW, Kim HG … Cho JY (2015) The dietary flavonoid Kaempferol mediates anti-inflammatory responses via the Src, Syk, IRAK1, and IRAK4 molecular targets. *Mediators of Inflammation* 2015:904142.

Kim S-H and Choi K-C (2013) Anti-cancer Effect and Underlying Mechanism(s) of Kaempferol, a Phytoestrogen, on the Regulation of Apoptosis in Diverse Cancer Cell Models. *Toxicological Research* 29:229–234.

Kirsh VA, Peters U, Mayne ST, Subar AF, Chatterjee N, Johnson CC and Hayes RB (2007) Prospective study of fruit and vegetable intake and risk of prostate cancer. *Journal of the National Cancer Institute* 99:1200–1209.

Kong L, Luo C, Li X, Zhou Y and He H (2013) The anti-inflammatory effect of kaempferol on early atherosclerosis in high cholesterol fed rabbits. *Lipids in Health and Disease* 12:115.

Kwon SH, Nam JI, Kim SH, Kim JH, Yoon J and Kim K (2009) Kaempferol and Quercetin , Essential Ingredients in Ginkgo biloba Extract, inhibit Interleukin-1 b -induced MUC5AC Gene Expression in Human Airway Epithelial Cells. *Phytotherapy Research* 23:1708–1712.

Labbé D, Provençal M, Lamy S, Boivin D, Gingras D and Béliveau R (2009) The flavonols quercetin, kaempferol, and myricetin inhibit hepatocyte growth factor-induced medulloblastoma cell migration. *The Journal of Nutrition* 139:646–652.

Lee E-J, Ji G-E and Sung M-K (2010c) Quercetin and kaempferol suppress immunoglobulin E-mediated allergic inflammation in RBL-2H3 and Caco-2 cells. *Inflammation Research* 59:847-854.

Lee HS, Cho HJ, Kwon GT, Han J and Park Y (2014) Kaempferol Downregulates Insulin-like Growth Factor-I Receptor and ErbB3 Signaling in HT-29 Human Colon Cancer Cells. *Journal of Cancer Prevention* 19:161–169.

Lee KM, Lee DE, Seo SK, Hwang MK, Heo Y-S, Lee KW and Lee HJ (2010a) Phosphatidylinositol 3-kinase, a novel target molecule for the inhibitory effects of kaempferol on neoplastic cell transformation. *Carcinogenesis* 31:1338–1343.

Lee KM, Lee KW, Jung SK, Lee EJ, Heo Y-S, Bode AM … Dong Z (2010b) Kaempferol inhibits UVB-induced COX-2 expression by suppressing Src kinase activity. *Biochemical Pharmacology* 80:2042–2049.

Leung HW-C, Lin C-J, Hour M-J, Yang W-H, Wang M-Y and Lee H-Z (2007) Kaempferol induces apoptosis in human lung non-small carcinoma cells accompanied by an induction of antioxidant enzymes. *Food and Chemical Toxicology : An International Journal Published for the British Industrial Biological Research Association* 45:2005–2013.

Li C, Zhao Y, Yang D, Yu Y, Guo H, Zhao Z … Yin X (2014) Inhibitory effects of kaempferol on the invasion of human breast carcinoma cells by downregulating the expression and activity of matrix metalloproteinase-9. *Biochemistry and Cell Biology* 93:16–27.

Li JJ, Westergaard C, Ghosh P and Colburn NH (1997) Inhibitors of both nuclear factor-kappaB and activator protein-1 activation block the neoplastic transformation response. *Cancer Research* 57:3569–3576.

Li W, Du B, Wang T, Wang S and Zhang J (2009) Kaempferol induces apoptosis in human HCT116 colon cancer cells via the Ataxia-Telangiectasia Mutated-p53 pathway with the involvement of p53 Upregulated Modulator of Apoptosis. *Chemico-Biological Interactions* 177:121–127.

Lim DY, Jeong Y, Tyner AL and Park JHY (2007) Induction of cell cycle arrest and apoptosis in HT-29 human colon cancer cells by the dietary compound luteolin. *American Journal of Physiology. Gastrointestinal and Liver Physiology* 292:G66–75.

Limtrakul P, Khantamat O and Pintha K (2005) Inhibition of P-glycoprotein function and expression by kaempferol and quercetin. *Journal of Chemotherapy* 17:86–95.

Lin CW, Chen PN, Chen MK, Yang WE, Tang CH, Yang SF and Hsieh YS (2013) Kaempferol reduces matrix metalloproteinase-2 expression by down-regulating ERK1/2 and the activator protein-1 signaling pathways in oral cancer cells. *PLoS ONE* 8:e80883.

López-Lázaro M (2006) Hypoxia-inducible factor 1 as a possible target for cancer chemoprevention. *Cancer Epidemiology, Biomarkers & Prevention : A Publication of the American Association for Cancer Research, Cosponsored by the American Society of Preventive Oncology* 15:2332–2335.

Lozanovski VJ, Houben P, Hinz U, Hackert T, Herr I and Schemmer P (2014) Pilot study evaluating broccoli sprouts in advanced pancreatic cancer (POUDER trial) - study protocol for a randomized controlled trial. *Trials* 15:1–8.

Luo C, Yang H, Tang C, Yao G, Kong L, He H and Zhou Y (2015) Kaempferol alleviates insulin resistance via hepatic IKK/NF-κB signal in type 2 diabetic rats. *International Immunopharmacology* 28:744–50.

Luo H, Daddysman MK, Rankin GO, Jiang B-H and Chen YC (2010a) Kaempferol enhances cisplatin's effect on ovarian cancer cells through promoting apoptosis caused by down regulation of cMyc. *Cancer Cell International* 10:16.

Luo H, Daddysman MK, Rankin GO, Jiang B-H and Chen YC (2010b) Kaempferol enhances cisplatin's effect on ovarian cancer cells through promoting apoptosis caused by down regulation of cMyc. *Cancer Cell International* 10:16.

Luo H, Jiang B, Li B, Li Z, Jiang BH and Chen YC (2012b) Kaempferol nanoparticles achieve strong and selective inhibition of ovarian cancer cell viability. *International Journal of Nanomedicine* 7:3951–3959.

Luo H, Rankin GO, Juliano N, Jiang B-H and Chen YC (2012a) Kaempferol inhibits VEGF expression and in vitro angiogenesis through a novel ERK-NFκB-cMyc-p21 pathway. *Food Chemistry* 130:321–328.

Luo H, Rankin GO, Li Z, Depriest L and Chen YC (2011) Kaempferol induces apoptosis in ovarian cancer cells through activating p53 in the intrinsic pathway. *Food Chemistry* 128:513–519.

Luo H, Rankin GO, Liu L, Daddysman MK, Jiang B-H and Chen YC (2009) Kaempferol inhibits angiogenesis and VEGF expression through both HIF dependent and independent pathways in human ovarian cancer cells. *Nutrition and Cancer* 61:554–563.

Luo J, Solimini NL and Elledge SJ (2009) Principles of cancer therapy: oncogene and non-oncogene addiction. *Cell* 136:823–837.

Marfe G, Tafani M, Indelicato M, Sinibaldi-Salimei P, Reali V, Pucci B … Russo MA (2009) Kaempferol induces apoptosis in two different cell lines via Akt inactivation, Bax and SIRT3 activation, and mitochondrial dysfunction. *Journal of Cellular Biochemistry* 106:643–650.

Matsuda H, Ninomiya K, Shimoda H and Yoshikawa M (2002) Hepatoprotective principles from the flowers of Tilia argentca (linden): structure requirements of tiliroside and mechanisms of action. *Bioorganic & Medicinal Chemistry* 10:707–712.

Mylonis I, Lakka A, Tsakalof A and Simos G (2010) The dietary flavonoid kaempferol effectively inhibits HIF-1 activity and hepatoma cancer cell viability under hypoxic conditions. *Biochemical and Biophysical Research Communications* 398:74–78.

Nadova S, Miadokova E and Cipak L (2007) Flavonoids potentiate the efficacy of cytarabine through modulation of drug-induced apoptosis. *Neoplasma* 54:202–206.

Nakamura Y, Chang CC, Mori T, Sato K, Ohtsuki K, Upham BL and Trosko JE (2005) Augmentation of differentiation and gap junction function by kaempferol in partially differentiated colon cancer cells. *Carcinogenesis* 26:665–671.

Navarro SL, Schwarz Y, Song X, Wang C, Chen C, Trudo SP … Lampe JW (2014) Cruciferous Vegetables Have Variable Effects on Biomarkers of Systemic Inflammation in a Randomized Controlled Trial in Healthy Young.

Ninomiya M, Nishida K, Tanaka K, Watanabe K and Koketsu M (2013) Structure-activity relationship studies of 5,7-dihydroxyflavones as naturally occurring inhibitors of cell proliferation in human leukemia HL-60 cells. *Journal of Natural Medicines* 67:460–467.

Nirmala P and Ramanathan M (2011) Effect of kaempferol on lipid peroxidation and antioxidant status in 1 , 2-dimethyl hydrazine induced colorectal carcinoma in rats. *European Journal of Pharmacology* 654:75–79.

Nomura M, He Z, Koyama I, Ma W-Y, Miyamoto K and Dong Z (2003) Involvement of the Akt/mTOR pathway on EGF-induced cell transformation. *Molecular Carcinogenesis* 38:25–32.

Nöthlings U, Murphy SP, Wilkens LR, Henderson BE and Kolonel LN (2007) Flavonols and pancreatic cancer risk: the multiethnic cohort study. *American Journal of Epidemiology* 166:924–931.

Oh H, Han N, Kim M, Kim H and Jeong H (2013) Evaluation of the effect of kaempferol in a murine allergic rhinitis model. *European Journal of Pharmacology* 718:48–56.

Park M-Y, Ji GE and Sung M-K (2012) Dietary kaempferol suppresses inflammation of dextran sulfate sodium-induced colitis in mice. *Digestive Diseases and Sciences* 57:355–363.

Petrick JL, Steck SE, Bradshaw PT, Trivers KF, Abrahamson PE, Engel LS … Gammon MD (2015) Dietary intake of flavonoids and oesophageal and gastric cancer: incidence and survival in the United States of America (USA). *British Journal of Cancer* 112:1291–1300.

Piao Y, Shin S-C and Choi J-S (2008) Effects of oral kaempferol on the pharmacokinetics of tamoxifen and one of its metabolites, 4-hydroxytamoxifen, after oral administration of tamoxifen to rats. *Biopharmaceutics & Drug Disposition* 29:245–249.

Rajendran P, Rengarajan T, Nandakumar N, Palaniswami R, Nishigaki Y and Nishigaki I (2014) Kaempferol, a potential cytostatic and cure for inflammatory disorders. *European Journal of Medicinal Chemistry* 86:103–112.

Ren H-J, Hao H-J, Shi Y-J, Meng X-M and Han Y-Q (2010) [Apoptosis-inducing effect of quercetin and kaempferol on human HL-60 cells and its mechanism]. *Zhongguo Shi Yan Xue Ye Xue Za Zhi/Zhongguo Bing Li*

Sheng Li Xue Hui = Journal of Experimental Hematology/Chinese Association of Pathophysiology 18:629–633.

Richman EL, Carroll PR and Chan JM (2012) Vegetable and fruit intake after diagnosis and risk of prostate cancer progression. *International Journal of Cancer. Journal International Du Cancer* 131:201–210.

Riso P, Vendrame S, Del Bo' C, Martini D, Martinetti A, Seregni E … Porrini M (2014) Effect of 10-day broccoli consumption on inflammatory status of young healthy smokers. *International Journal of Food Sciences and Nutrition* 65:106–111.

Ruiz E, Padilla E, Redondo S, Gordillo-Moscoso A and Tejerina T (2006) Kaempferol inhibits apoptosis in vascular smooth muscle induced by a component of oxidized LDL. *European Journal of Pharmacology* 529:79–83.

Sak K (2014) Cytotoxicity of dietary flavonoids on different human cancer types. *Pharmacognosy Reviews* 8:122–146.

Schindler R and Mentlein R (2006) Flavonoids and vitamin E reduce the release of the angiogenic peptide vascular endothelial growth factor from human tumor cells. *The Journal of Nutrition* 136:1477–1482.

Schinkel AH and Jonker JW (2003) Mammalian drug efflux transporters of the ATP binding cassette (ABC) family: an overview. *Advanced Drug Delivery Reviews* 55:3–29.

Semenza GL (2006) Development of novel therapeutic strategies that target HIF-1. *Expert Opinion on Therapeutic Targets* 10:267–280.

Semenza GL (2007) Evaluation of HIF-1 inhibitors as anticancer agents. *Drug Discovery Today* 12:853–859.

Sharma V, Joseph C, Ghosh S, Agarwal A, Mishra MK and Sen E (2007) Kaempferol induces apoptosis in glioblastoma cells through oxidative stress. *Molecular Cancer Therapeutics* 6:2544–2553.

Shih T, Young T, Lee H, Hsieh C and Hu OY (2013) Protective Effects of Kaempferol on Isoniazid- and Rifampicin-Induced Hepatotoxicity. *AAPS Journal* 15:753-762.

Srinivas Raghavan B, Kondath S, Anantanarayanan R and Rajaram R (2015) Kaempferol mediated synthesis of gold nanoparticles and their cytotoxic effects on MCF-7 cancer cell line. *Process Biochemistry* 50:1966–1976.

Steevens J, Schouten LJ, Goldbohm RA and Van Den Brandt PA (2011) Vegetables and fruits consumption and risk of esophageal and gastric cancer subtypes in the Netherlands Cohort Study. *International Journal of Cancer* 129:2681–2693.

Szliszka E, Helewski KJ, Mizgala E and Krol W (2011) The dietary flavonol fisetin enhances the apoptosis-inducing potential of TRAIL in prostate cancer cells. *International Journal of Oncology* 39:771–779.

Tang X-L, Liu J-X, Dong W, Li P, Li L, Hou J-C … Ren J-G (2015) Protective effect of kaempferol on LPS plus ATP-induced inflammatory response in cardiac fibroblasts. *Inflammation* 38:94–101.

Thomas R, Williams M, Sharma H, Chaudry A and Bellamy P (2014). A double-blind , placebo-controlled randomised trial evaluating the effect of a polyphenol-rich whole food supplement on PSA progression in men with prostate cancer — the UK NCRN Pomi-T study. *Prostate Cancer and Prostatic Disease* 17:180–186.

Thorburn A (2004) Death receptor-induced cell killing. *Cellular Signalling* 16:139–144.

Vogelstein B and Kinzler KW (2004) Cancer genes and the pathways they control. *Nature Medicine* 10:789–799.

Wang L, Lee I-M, Zhang SM, Blumberg JB, Buring JE and Sesso HD (2009) Dietary intake of selected flavonols, flavones, and flavonoid-rich foods and risk of cancer in middle-aged and older women. *The American Journal of Clinical Nutrition* 89:905–912.

Wang M, Sun J, Jiang Z, Xie W and Zhang X (2015) Hepatoprotective effect of kaempferol against alcoholic liver injury in mice. *The American Journal of Chinese Medicine* 43:241–254.

Weng CJ and Yen GC (2012) Flavonoids, a ubiquitous dietary phenolic subclass, exert extensive in vitro anti-invasive and in vivo anti-metastatic activities. *Cancer and Metastasis Reviews* 31:323–351.

WHO (2014) *World Cancer Report 2014.* (B. W. Stewart and C. P. Wild, Eds.). IARC.

Xiao J, Sun G-B, Sun B, Wu Y, He L, Wang X … Sun X-B (2012) Kaempferol protects against doxorubicin-induced cardiotoxicity in vivo and in vitro. *Toxicology* 292:53–62.

Xie F, Su M, Qiu W, Zhang M, Guo Z, Su B … Zhou L (2013) Kaempferol promotes apoptosis in human bladder cancer cells by inducing the tumor suppressor, PTEN. *International Journal of Molecular Sciences* 14:21215–21226.

Xu W, Liu J, Li C, Wu H-Z and Liu Y-W (2008) Kaempferol-7-O-beta-D-glucoside (KG) isolated from Smilax china L. rhizome induces G2/M phase arrest and apoptosis on HeLa cells in a p53-independent manner. *Cancer Letters* 264:229–240.

Yang E-J, Kim G-S, Jun M and Song K-S (2014) Kaempferol attenuates the glutamate-induced oxidative stress in mouse-derived hippocampal neuronal HT22 cells. *Food & Function* 5:1395–1402.

Yasukawa K, Takido M, Takeuchi M, Sato Y, Nitta K and Nakagawa S (1990) Inhibitory effects of flavonol glycosides on 12-O-tetradecanoylphorbol-13-acetate-induced tumor promotion. *Chemical & Pharmaceutical Bulletin* 38:774–776.

Yoo S, Cho SJ and Cho Y (2015) Molecular Targeting of ERKs/RSK2 Signaling Axis in Cancer Prevention. *Journal of Cancer Prevention* 20:165–171.

Yoshida T, Konishi M, Horinaka M, Yasuda T, Goda AE, Taniguchi H … Sakai T (2008) Kaempferol sensitizes colon cancer cells to TRAIL-induced apoptosis. *Biochemical and Biophysical Research Communications* 375:129–133.

Yunnus M, Mahat A, Kulkarni NM, Vishwakarma SL, Khan FR, Mohammad S … Patil BM (2010) Modulation of the cyclooxygenase pathway via inhibition of nitric oxide production contributes to the anti-in fl ammatory activity of kaempferol. *European Journal of Pharmacology* 642:169–176.

Zhang Y, Chen AY, Li M, Chen C and Yao Q (2008) Ginkgo biloba extract kaempferol inhibits cell proliferation and induces apoptosis in pancreatic cancer cells. *The Journal of Surgical Research* 148:17 23.

Zhong H, De Marzo AM, Laughner E, Lim M, Hilton DA, Zagzag D … Simons JW (1999) Overexpression of hypoxia-inducible factor 1alpha in common human cancers and their metastases. *Cancer Research* 59:830–5835.

In: Kaempferol ISBN: 978-1-63485-828-1
Editors: T.Garde-Cerdán and A.G. Diago ©2016 Nova Science Publishers, Inc.

Chapter 4

KAEMPFEROL: REVIEW ON NATURAL SOURCES AND BIOAVAILABILITY

Muhammad Jahangir Hossen[1,2,*], Md Bashir Uddin[3,4], Syed Sayeem Uddin Ahmed[5], Zhi-Ling Yu[6] and Jae Youl Cho[2]

[1]Department of Animal Science, Patuakhali Science and Technology University, Dumki, Patuakhali, Bangladesh
[2]Department of Genetic Engineering, Sungkyunkwan University, Suwon, Republic of Korea
[3]College of Veterinary Medicine, Chungnam National University, Daejeon, Republic of Korea
[4]Department of Medicine and
[5]Department of Epidemiology and Public Health, Sylhet Agricultural University, Sylhet, Bangladesh
[6]School of Chinese Medicine, Hong Kong Baptist University, Kowloon Tong, Hong Kong, China

[*] Corresponding author; Associate Professor, Department of Animal Science, Patuakhali Science & Technology University, Dumki, Patuakhali, 88-8602, Bangladesh. Tel.: +88-04427-56014 Ext.273 (off); Fax: +88-4427-56009. E-mail addresses: jhossen_pstu@yahoo.com.

ABSTRACT

Recently, researchers have focused on dietary products due to the many health benefits and remedial potential of pharmacologically active compounds such as flavonoids. Kaempferol (3,5,7-trihydroxy-2-(4-hydroxyphenyl)-4H-1-benzopyran-4-one) is a flavonoid linked to diverse glycoside moieties and is extremely plentiful in most edible plants such as tea, fruits and vegetables including *Allium cepa* (onion), *Camellia sinensis* (tea), *Citrus paradisi* (grapefruit), *Fragaria vesca* (strawberry), *Lactuca sativa* (lettuce), and *Morinda citrifolia* (Indian mulberry) as well as in medicinal plants such as *Acacia nilotica* (L), *Aloe vera* (L.), *Crocus sativus* (L.), *Euphorbia pekinensis* Rupr., *Ginkgo biloba* (L.), *Hypericum perforatum* (L.), *Phyllanthus acidus* (L.), *Ribes nigrum* (L.), *Rosmarinus officinalis* (L.), *Cerbera manghas*, and *Persicaria chinensis* (L.). Numerous epidemiological investigations determined that kaempferol rich foods ameliorate several disorders including cancer and cardiovascular disease. Molecular mechanistic studies report that kaempferol modulates a number of key elements in the cellular signal transduction pathway linked to apoptosis, angiogenesis, inflammation, and metastasis. These studies show that kaempferol has lower toxicity in comparison to standard chemotherapy drugs. Kaempferol demonstrates poor oral bioavailability and is commonly metabolized into different forms such as methyl, sulfate or glucuronide, however its combination with other anticancer agents enhances its anticancer properties. This chapter provides several natural sources of kaempferol with its pharmacokinetics (oral availability) and safety. This review shows the dietary wealth of kaempferol with its bioavailability and may contribute to the further use of this flavonoid as a prospective novel candidate for future drug development.

Keywords: anti-cancer, antiinflammation, bioavailability, dietary sources, flavonoid, kaempferol

1. INTRODUCTION

Flavonoids are diphenylpropane structures widely distributed in the plant kingdom and are common constituents of fruits, vegetables and some beverages. Ingestion of flavonoid containing foods reduces the risk of developing chronic diseases such as cancer and cardiovascular disease (Maron, 2004; Neuhouser, 2004) and certain flavonoids may be useful in the treatment of several diseases.(Li et al., 2007; Lopez-Lazaro, 2009).

The flavonoid kaempferol (3,5,7-trihydroxy-2-(4-hydroxyphenyl)-4H-1-benzopyran-4-one) is a yellow compound with a molecular weight of 286.2 g/mol and is commonly found in plant derived foods and medicinal plants. Although there are over three thousand articles in PubMED reporting the isolation and biological properties of kaempferol, there is no summary update with information regarding its natural sources and bioavailability. In this review, we benchmark kaempferol and its glycosides which contain natural resources, which may help develop this flavonoid as a possible agent for the prevention and treatment of some diseases.

2. Natural Sources of Kaempferol

Kaempferol is extremely plentiful in most edible plants such as tea, fruits and vegetables including *Allium cepa* (onion), *Camellia sinensis* (tea), *Citrus paradisi* (grapefruit), *Fragaria vesca* (strawberry), *Lactuca sativa* (lettuce), and *Morinda citrifolia* (Indian mulberry) as well as in medicinal plants such as *Acacia nilotica* (L), *Aloe vera* (L.), *Crocus sativus* (L.), *Euphorbia pekinensis* Rupr., *Ginkgo biloba* (L.), *Hypericum perforatum* (L.), *Phyllanthus acidus* (L.), *Ribes nigrum* (L.), *Rosmarinus officinalis* (L.), *Cerbera manghas*, and *Persicaria chinensis* (L.). The plant species in which kaempferol and/ or kaempferol glycosides have been identified are summarized in Table 1 in alphabetic order by family with references.

Table 1. List of plant species containing kaempferol and/or its glycosides

Family name	Name of Species	References
Acanthacea	*Strobilanthes crispus*	(Ghasemzadeh et al., 2015a)
Actinidiaceae	***Actinidia valvata***	(Xin et al., 2011)
Acoraceae	***Acorus gramineus***	(Park et al., 2011)
Alangiaceae	*Alangium salviifolium*	(Hung et al., 2009)
Alliaceae	*Allium cepa*	(Rodríguez Galdón et al., 2008)
	Allium hirtifolium	(Barile et al., 2005)
	Allium neapolitanum	(Carotenuto et al., 1997)
	Allium porrum	(Fattorusso et al., 2001)
	Allium triquetrum	(Corea et al., 2003)

Table 1. (Continued)

Family name	Name of Species	References
	Allium ursinum	(Carotenuto et al., 1996)
	Allium victorialis	(Lee et al., 2001)
	Alternanthera tenella	(Salvador et al., 2006)
	Amaranthus spinosus	(Stintzing et al., 2004)
Altingiaceae	*Liquidambar formosana*	(Liao et al., 2014)
Amaranthaceae	*Aerva javanica*	(Mussadiq et al., 2013)
Amaryllidaceae	*Allium cepa*	(Ikechukwu and Ifeanyi, 2015)
	Allium ursinum	(Oszmianski et al., 2012)
Anacardiaceae	*Pistacia chinensis*	(Zhang et al., 2016)
	Pistacia vera	(Tomaino et al., 2010)
	Rhus verniciflua	(Lin et al., 2010)
Annonaceae	*Annona purpurea*	(Chang et al., 1998)
	Ellipeiopsis cherrevensis	(Wirasathien et al., 2006)
	Uvaria tonkinensis	(Liu et al., 2005)
	Annona crassiflora	(Rocha et al., 2015)
Apiaceae	*Angelica shikokiana*	(Mira et al., 2015)
	Ammi majus	(Singab, 1998)
	Bunium persicum	(Sharififar et al., 2010)
	Bupleurum flavum	(Pistelli et al., 2005)
	Centella asiatica	(Satake et al., 2007; Suntornsuk and Anurukvorakun, 2005)
	Foeniculum vulgare	(Parejo et al., 2004)
	Pleurospernum franchetianum	(Luo et al., 2002)
Apocynaceae	*Apocynum venetum*	(Grundmann et al., 2009; Xiong et al., 2000)
	Forsteronia refracta	(Xu et al., 2006)
	Thevetia peruviana	(Abe et al., 1995)
	Dipladenia martiana	(de Carvalho et al., 2001)
	Echites hirsuta	(Chien et al., 1979)
	Gynanchum paniculatum	(Fu et al., 2015)
Asclepiadaceae	*Asclepias incarnata*	(Sikorska, 2003)
	Asclepias syriaca	(Sikorska et al., 2001)
	Baseonema acuminatum	(De Leo et al., 2004)

Family name	Name of Species	References
	Solenostemma argel	(Kamel et al., 2000)
	Cynanchum acutum	(Anwar et al., 2007)
	Cynanchum chinense	(Liu et al., 2006)
	Gymnema sylvestre	(Liu et al., 2004)
Asphodelaceae	*Aloe vera*	(Keyhanian and Stahl-Biskup, 2007)
Aspleniaceae	*Asplenium prolongatum*	(Mizuno et al., 1990)
	Camposorus sibiricus	(Li et al., 2006b)
	Asplenium trichomanes	(Dall'Acqua et al., 2009)
Aspidiaceae	*Dryopteris crassirhizoma*	(Min et al., 2001)
Asteraceae	*Alomia myriadenia*	(Scio et al., 2003)
	Anaphalis aureopunctata	(Wu et al., 2003)
	Carthamus lanatus	(Taskova et al., 2003)
	Carthamus tinctorius	(Ahmed et al., 2000)
	Centaurea hierapolitana	(Karamenderes et al., 2007)
	Chromolaena odorata	(Ling et al., 2007)
	Chuquiraga spinosa	(Landa et al., 2009)
	Cichorium endivia	(DuPont et al., 2000)
	Senecio scandens	(Wang et al., 2010)
	Silphium perfoliatum	(El-Sayed et al., 2002)
	Solidago altissima	(Wu et al., 2007)
	Solidago virga-aurea	(Choi et al., 2004)
	Cirsium rivulare	(Nazaruk and Jakoniuk, 2005)
	Conyza aegyptiaca	(Mahmoud et al., 2009)
	Conyza filaginoides	(Calzada et al., 2001)
	Crassocephalum crepidioides	(Aniya et al., 2005)
	Grindelia robusta	(Krenn et al., 2009)
	Helichrysum italicum	(Sala et al., 2003)
	Vernonia ferruginea	(Malafronte et al., 2009)
	Vernonia travancorica	(Seetharaman and Petrus, 2004)
Basellaceae	*Ullucus tuberosus*	(Dini et al., 1991)
Balsaminaceae	*Impatiens balsamina*	(Lim et al., 2007)
	Impatiens textori	(Ueda et al., 2003)
Berberidaceae	*Dysosma versipellis*	(Jiang et al., 2007)
	Epimedium pubescens	(Zhang et al., 2013)
	Epimedium sagittatum	(Wang et al., 2007)

Table 1. (Continued)

Family name	Name of Species	References
	Epimedium wushanense	(Li et al., 2012)
Blechnaceae	*Blechnum novae-zelandiae*	(Swinny, 2001)
	Stenochlaena palustris	(Liu et al., 1999)
Bigononiaceae	*Crescentia alata*	(Autore et al., 2001)
	Jacaranda acutifolia	(Mostafa et al., 2015)
	Oroxylum indicum	(Wei et al., 2013)
Boraginaceae	*Thyrocsrpus glochiditus*	(Luo et al., 2006)
	Borago officinalis	(Samy et al., 2015)
	Arnebia purpurea	(Yuzbasioglu et al., 2015)
Brassicaceae	*Arabidopsis mutanats*	(Ryan et al., 2001)
	Arabidopsis thaliana	(Hashiguchi et al., 2013; Veit and Pauli, 1999)
	Brassica spp	(Nielsen et al., 1993)
	Brassica campestris	(Harbaum et al., 2007)
	Brassica juncea	(Kim et al., 2002)
	Brassica oleracea	(Olsen et al., 2009)
	Brassica rapa	(Rochfort et al., 2006)
	Bunias orientalis	(Bennett et al., 2006)
	Cardamine tangutorum	(Feng et al., 2012)
	Diplotaxis erucoides	(Bennett et al., 2006)
	Diplotaxis harra	(Kassem et al., 2013)
	Diplotaxis tenuifolia	(Bennett et al., 2006)
Brassicaceae	*Draba nemorosa*	(Moon et al., 2010)
	Heliophila coronopifolia	(Saito et al., 2011)
	Rapistrum rugosum	(Al-Taweel et al., 2012)
Burseraceae	*Canarium pimela*	(Lv et al., 2014)
	Eruca sativa	(Bennett et al., 2006)
Caricaceae	*Carica papya*	(Gogna et al., 2015)
Cactaceae	*Cephalocereus senilis*	(Qin et al., 1994)
	Opuntia dillenii	(Qiu et al., 2002)
	Opuntia ficus indica	(Lee et al., 2003)
Caesalpiniaceae	*Pterogyne nitens*	(Regasini et al., 2008)
Campanulaceae	*Campanula alliariifolia*	(Dumlu et al., 2008)
	Campanula barbata	(Cuendet et al., 2001)
Cannabaceae	*Cannabis sativa*	(Ross et al., 2005)
Canellaceae	*Warburgia stuhlmannii*	(Manguro et al., 2003a)

Family name	Name of Species	References
	Warburgia ugandensis	(Manguro et al., 2003b)
Capparaceae	*Capparis spinosa*	(Bonina et al., 2002)
Caprifoliaceae	*Lonicera japonica*	(Choi et al., 2007)
	Lonicera macranthoides	(Wu et al., 2012a)
	Sambucus nigra	(Schmitzer et al., 2010)
Caryophyllaceae	*Dianthus barbatus*	(Cordell et al., 1976)
Celastraceae	*Celastrus hindsii*	(Ly et al., 2006)
	Celastrus tatrinovii	(Rzadkowska-Bodalska et al., 1974)
	Celastrus orbiculatus	(Yu et al., 2014)
	Euonymus alatus	(Fang et al., 2008)
	Maytenus aquifolium	(Vilegas et al., 1999)
	Maytenus ilicifolia	(Leite et al., 2001)
	Euonymus fortunei	(Yan et al., 2015)
Cephalotaxaceae	*Cephalotaxus koreana*	(Yoon et al., 2007)
Chenopodiaceae	*Chenopodium murale*	(Gohar et al., 2000)
	Suaeda maritima	(Abd El-Latif et al., 2014)
Chrysobalanaceae	*Licania licaniaeflora*	(Bonina et al., 2002)
	Chenopodium quinoa	(De Simone et al., 1990)
Cistaceae	*Cistus ladanifer*	(Sosa et al., 2004)
	Cistus laurifolius	(Küpeli et al., 2006)
Clusiaceae	*Hypericum brasiliense*	(Rocha et al., 1995)
	Hypericum perforatum	(Odabas et al., 2010)
	Vismia lorenti	(Nguemeving et al., 2006)
	Clusia nemorosa	(Ferreira et al., 2015)
Combretaceae	*Terminalia myriocarpa*	(Marzouk et al., 2002)
Convolvulaceae	*Argyreia speciosa*	(Habbu et al., 2009)
	Cuscuta australis	(Ye et al., 2005)
	Cuscuta chinensis	(Umehara et al., 2004)
	Evolvulus alsinoides	(Kumar et al., 2010)
Cornaceae	*Cornus kousa*	(Vareed et al., 2007)
Corylaceae	*Corylus avellana*	(Amaral et al., 2005)
Crassulaceae	*Kalanchoe pinnata*	(Muzitano et al., 2006)
	Rhodolia rosea	(Jeong et al., 2009a)
	Rhodolia sachalinensis	(Song et al., 2003)
	Orostachys japonicus	(Je Ma et al., 2009)
	Sedum dendroideum	(De Melo et al., 2005)
	Sedum sarmentosum	(Oh et al., 2004a)

Table 1. (Continued)

Family name	Name of Species	References
	Sempervivum tectorum	(Stojković et al., 2015)
Crypteroniaceae	*Crypteronia paniculata*	(Deng et al., 2002)
Cucurbitaceae	*Gynostemma cardiospermum*	(Yin et al., 2006)
	Neoalsomitra integrifoliola	(Su et al., 2012)
	Siraitia grosvenori	(Li et al., 2006a)
Cyatheaceae	*Cyathea phalerata*	(Cazarolli et al., 2006)
Daphiniphyllaceae	*Daphiniphyllum calycinum*	(Gamez et al., 1998)
Dennstaedtiaceae	*Dennstaedtia scabra*	(Li et al., 2009a)
Dioscoreaceae	*Dioscorea bulbifera*	(Gao et al., 2002)
Dipsacaceae	*Scabiosa hymettia*	(Christopoulou et al., 2008)
Ebenaceae	*Diospyros crassiflora*	(Akak et al., 2010)
	Diospyros kaki	(Chen et al., 2002)
	Diospyros lotus	(Loizzo et al., 2009)
Elaegnaceae	*Hippophae rhamnoides*	(Fatima et al., 2015; Sharma et al., 2007)
Elatinaceae	*Bergia ammanniodes*	(Ezzat et al., 2015)
Epacridaceae	*Ardisia japonica*	(Li et al., 2005)
Equisetaceae	*Equisetum arvense*	(Oh et al., 2004b)
	Equisetum debile	(Keyhanian and Stahl-Biskup, 2007)
	Equisetum giganteum	(Francescato et al., 2013)
	Equisetum myriochaetum	(Cetto et al., 2000)
	Equisetum palustre	(Gurbuz et al., 2009)
Ericaceae	*Calluna vulgaris*	(Orhan et al., 2007)
	Vaccinium bracteatum	(Lv et al., 2012)
Eucommiaceae	*Eucommia ulmoides*	(Yang et al., 2014)
Euphorbiaceae	*Acalypha hispida*	(Adesina et al., 2000)
	Cnidoscolus aconitifolius	(Kuti and Konuru, 2004)
	Cnidoscolus chayamansa	
	Croton cajucara	(Maciel et al., 2000)
	Croton gossypifolius	(Quintyne-Walcott et al., 2007)
	Elateriospermum tapos	(Pattamadilok and Suttisri, 2008)
	Excoecaria acerifclia	(Hu et al., 2011)

Family name	**Name of Species**	**References**
	Euphorbia aleppica	(Öksüz et al., 1996)
	Euphorbia dracunculoides	(Wang et al., 2014)
	Euphorbia hirta	(Yi et al., 2012)
	Euphorbia pekinensis	(Hwang et al., 2001)
	Euphorbia petiolata	(Nazemiyeh et al., 2010)
	Euphorbia lunulata	(Liu et al., 2015)
	Euphorbia sikkimensis	(Yang et al., 2013)
	Manihot esculenta	(Prawat et al., 1995)
	Pedilanthus tithymaloides	(Abreu et al., 2008)
	Sapium sebiferum	(Gao et al., 2015)
	Sauropus androgynus	(Yang et al., 2008)
	Excoecaria venenata	(Ning et al., 2015)
Fabaceae	*Astragalus monspessulanus*	(Krasteva et al., 2015)
	Astragalus abyssinicus	(El Dib et al., 2015)
	Delonix elata	(Al-Taweel et al., 2015)
	Desmodium caudatum	(Wu et al., 2012b)
	Flemingia philippinensis	(Fu et al., 2013)
Fabaceae	*Medicago lupulina*	(Kicel and Olszewska, 2015)
	Glycine max	(Li et al., 2015)
	Leucaena leucocephala	(Hassan et al., 2013)
	Securigera securidaca	(Behbahani et al., 2014)
	Trifolium repens	(Kicel and Wolbiś, 2012)
Fagaceae	*Quercus dentata*	(Meng et al., 2001)
	Pasania dodoniifolia	(Chang and Lee, 2015)
Foaceae	*Cynodon dactylon*	(Muthukrishnan et al., 2015)
Frankeniaceae	*Frankenia laevis*	(Hussein, 2004)
Gentianaceae	*Blackstonia perfoliata*	(Kaouadji et al., 1990)
Geraniaceae	*Geranium bellum*	(Gayosso-De-Lucio et al., 2010)
	Geranium carolinianum	(Li et al., 2008)
	Geranium pratense	(Küpeli et al., 2007)
	Geranium potentillaefolium	(Gayosso-De-Lucio et al., 2010)
	Pelargonium quercifolium	(Williams et al., 1997)
Ginkgoaceae	*Ginkgo biloba*	(Kwon et al., 2009)

Table 1. (Continued)

Family name	Name of Species	References
Hippocastanaceae	*Aesculus hippocastanum*	(Dudek-Makuch and Matławska, 2011)
Hydrangeaceae	*Philadelphus tenuifolius*	(Grančai et al., 2014)
Hypericaceae	*Hypericum hengshanense*	(Wang and Xu, 2013)
Illiciaceae	*Illicium simonsii*	(Liu et al., 2011a)
Iridaceae	*Crocus antalynensis*	(Nørbæk and Kondo, 1999)
	Crocus speciosus	
	Crocus sativus	(Li et al., 2004; Mokhtari-Zaer et al., 2015)
Juglandaceae	*Cyclocarya paliurus*	(Xie et al., 2015)
	Juglans regia	(Qureshi et al., 2014)
Lamiaceae	*Callicarpa peii*	(Hu et al., 2013)
	Dracocephalum peregrinum	(Dai et al., 2008)
	Lagopsis supina	(Zhang et al., 2015a)
	Origanum dictamnus	(Chatzopoulou et al., 2010)
	Pogostemon cablin	(Ruan et al., 2013)
	Rosmarinus officinalis	(Bai et al., 2010)
Lauraceae	*Cinnamomum osmophloeum*	(Fang et al., 2005)
	Lindera aggregata	(Xiao et al., 2011)
Lecythidaceae	*Planchonia grandis*	(Crublet et al., 2003)
Leguminosae	*Astragalus caprinus*	(Semmar et al., 2002)
	Astragalus shikokianus	(Yahara et al., 2000)
	Bauhinia forficata	(Filomeni et al., 2012)
	Bauhinia malabarica	(Kaewamatawong et al., 2008)
	Bauhinia megalandra	(Rodríguez et al., 2010)
	Bauhinia microstachya	(Meyre-Silva et al., 2001)
	Bauhinia variegata	(Rao et al., 2008)
	Canavalia gladiata	(Murakami et al., 2000)
	Cassia alata	(Moriyama et al., 2003)
	Cassia angustifolia	(Terreaux et al., 2002)
	Cassia nodosa	(Kumar et al., 2006)
	Cassia siamea	(Ntandou et al., 2010)
	Clitoria ternatea	(Kazuma et al., 2003)
	Derris trifoliata	(Xu et al., 2009)

Family name	Name of Species	References
Leguminosae	*Dorycnium rectum*	(Moreno et al., 2002)
	Galega officinalis	(Champavier et al., 2000)
	Glycine max	(Ho et al., 2002)
	Glycyrrhiza spp	(Hatano et al., 1989)
	Gueldenstaedtia stenophylla	(Wang et al., 2012)
	Indigofera suffruticosa	(Calvo et al., 2011)
	Indigofera truxillensis	
	Oxytropis falcate	(Jiang et al., 2008)
	Securigera securidaca	(Ali et al., 1998)
	Tadehagi triquetrum	(Xiang et al., 2005)
	Trifolium alexandrinum	(Sharaf, 2008)
	Acacia nilotica	(Singh et al., 2008)
Loranthaceae	*Dendrophthoe falcata*	(Mallavadhani et al., 2006)
Liliaceae	*Lilium candidum*	(Vachálková et al., 1999)
	Lilium longiflorum	(Francis et al., 2004)
	Lilium pumilum	(Obmann et al., 2010)
Lythraceae	*Ammania auriculata*	(Nawwar et al., 2015)
	Cuphea pinetorum	(Calzada, 2005)
Malpighiaceae	*Memorialis hirta*	(Lei et al., 2012)
Marsileaceae	*Marsilea quadrifolia*	(Zhang et al., 2015b)
Malvaceae	*Althaea rosea*	(Papiez et al., 2001)
	Hibiscus sabdariffa	(Zhen et al., 2016)
	Urena lobata	(Jia et al., 2011)
Melastomataceae	*Melastoma malabathricum*	(Wong et al., 2012)
	Melastoma dodecandrum	(Cheng et al., 2014)
Meliaceae	*Amoora cucullata*	(Abdelfattah et al., 2010)
Moraceae	*Cudrania tricuspidata*	(Zou et al., 2004)
	Cudrania cochinchinensis	(Zhou et al., 2013)
	Ficus benjamina	(Yarmolinsky et al., 2012)
	Ficus pandurata	(Ramadan et al., 2009)
Myrsinaceae	*Ardisia colorata*	(Sumino et al., 2002)
Myrtaceae	*Callistemon lanceolatus*	(Mahmoud et al., 2002)
	Eucalyptus spp	(Amakura et al., 2009)
	Eucalyptus occidentalis	(Benyahia et al., 2004)
	Psidium guajava	(Chen and Nunez, 2010; Shao et al., 2014)
	Syzygium aromaticum	

Table 1. (Continued)

Family name	Name of Species	References
Nepenthaceae	*Nepenthes gracilis*	(Aung et al., 2002)
Nelumbonaceae	*Nelumbo nucifera*	(Lee et al., 2015; Zhao et al., 2013)
Oleaceae	*Chionanthus retusus*	(Kwak et al., 2009)
	Olea europaea	(De Laurentis et al., 1997)
	Olax manni	(Okoye et al., 2015)
	Ximenia caffra	(Zhen et al., 2015)
Ophioglossaceae	*Ophioglossum petolatum*	(Lin et al., 2005)
	Ophioglossum vulgatum	(Clericuzio et al., 2012)
Orchidaceae	*Anoectochilus roxburghii*	(Xiao et al., 2014)
	Goodyera schlechtendaliana	(Du et al., 2000)
Orobanchaceae	*Pedicularis densispica*	(Chu et al., 2011)
Papaveraceae	*Hylomecon vernalis*	(Lee et al., 2012)
Pedaliaceae	*Sesamum indicum*	(Dat et al., 2016)
Phyllanthaceae	*Phyllanthus acidus*	(Hossen et al., 2015b)
Primulaceae	*Androsace umbellata*	(Lei et al., 2011)
Pinaaceae	*Pseudotsuga menziesii*	(Krauze-Baranowska et al., 2013)
Platanaceae	*Platanus acerifolia*	(Zuo et al., 2015)
Polygonaceae	*Antenoron filiforme*	(Zhao et al., 2011)
	Persicaria chinensis	(Hossen et al., 2015a)
	Polygonum tinctorium	(Kataoka et al., 2001)
	Polygonum viscosum	(Das and Ganapaty, 2015)
	Polygonum amplexicaule	(Mirza and Ahmad, 2015)
Polypodiaceae	*Drynaria fortunei*	(Wang et al., 2008)
	Calligonum comosum	(Badria et al., 2007)
	Phymatopteris hastate	(Duan et al., 2012)
	Pyrrosia calvata	(Chen et al., 2015)
	Drynaria bonii	(Trinh et al., 2015)
Primulaceae	*Lysimachia clethroides*	(Liang et al., 2015)
	Primula sieboidii	(Hashimoto et al., 2015)
	Lysimachia circaeoides	(Wei et al., 2013)
Punicaceae	*Punica granatum*	(Van Elswijk et al., 2004)
Ranunculaceae	*Aconitum tanguticum*	(Xu et al., 2013)
	Consolida oliveriana	(Marín et al., 2009)

Family name	Name of Species	References
	Delphinium gracile	(Diaz and Herz, 2010)
	Helleborus niger	(Vitalini et al., 2011)
	Pulsatilla koreana	(Liu et al., 2012)
	Nigella glandulifera	(Liu et al., 2011b)
Rhamnaceae	*Berchemia floribunda*	(Wang et al., 2006)
	Colubrina asiatica	(Lee et al., 2000)
	Ficaria verna	(Tomczyk et al., 2002)
Rhamnaceae	*Rhamnus nakahari*	(Wei et al., 2001)
	Rhamnus nipalensis	(Singh et al., 2008)
	Rhamnus petiolaris	(Özipek et al., 1994)
	Rhamnus procumbens	(Goel et al., 1988)
Rhizophoraceae	*Cassipourea gummiflua*	(Drewes and Taylor, 1994)
Rosaceae	*Amelanchier alnifolia*	(Lavola et al., 2012)
	Eriobotrya japonica	(Benyahia et al., 2004; Rashed and Butnariu, 2014)
	Exochorda racemosa	(Zhang et al., 2011)
	Fragaria ananassa	(Aaby et al., 2007) Aaby et al., 2007)
	Potentilla multicaulis	(Jia et al., 2013)
	Rosa spp	(Nowak and Gawlik-Dziki, 2007)
	Rosa hybrids	(Suntornsuk and Anurukvorakun, 2005)
	Rosa sericea	(Li et al., 2013)
Rubiaceae	*Diodia teres*	(Lee et al., 2004)
	Gardenia jasminoides	(Song et al., 2013)
	Hedyotis diffusa	(Kim et al., 2001)
	Morinda citrifolia	(Deng et al., 2007)
	Psychotria straminea	(Fu et al., 2015)
	Oldenlandia diffusa	(Zhang and Sun, 2014)
Rutaceae	*Citrus ladinifier*	(Berhow et al., 1994)
Sapindaceae	*Koelreuteria paniculata*	(Qu et al., 2011)
Salicaceae	*Populus davidiana*	(Zhang et al., 2006)
	Oncoba spinosa	(Djouossi et al., 2015)
Saxifragaceae	*Bergenia stracheyi*	(Srivastava et al., 2015)
Scrophulariaceae	*Verbascum blattaria*	(Youn et al., 2015)
Simaroubaceae	*Simarouba versicolor*	(Arriaga et al., 2002)
Solanaceae	*Capsicum annuum*	(Mokhtar et al., 2015)

Table 1. (Continued)

Family name	Name of Species	References
	Datura suaveolens	(Sajeli Begum et al., 2006)
	Solanum nigrum	(Huang et al., 2010;
	Solanum lycopersicum	(Ferreres et al., 2010)
	Solanum macaonense	(Lee et al., 2015)
	Solanum schimperianum	(Al-Oqail et al., 2012)
Staphyleaceae	*Euscaphis japonica*	(Lee et al., 2007)
Sterculiaceae	*Theobroma grandiflorum*	(Yang et al., 2003)
Tamaricaceae	*Tamarix chinensis*	(Zhao et al., 2014)
Tiliaceae	*Tilia tomentosa*	(Viola et al., 1994)
Theaceae	*Camellia sinensis*	(Lee et al., 2008)
Thymelaeceae	*Gnidia involucrata*	(Ferrari et al., 2000)
Tofieldiaaceae	*Triantha japonica*	(Iwashina et al., 2013)
	Tofieldia nuda	
	Wikstroemia indica	(Sun et al., 2015)
Ulmaceae	*Zelkova oregoniana*	(Niklas and Giannasi, 1977)
Vahiliaceae	*Vahlia capensis*	(Majinda et al., 1997)
Velloziaceae	*Barbacenia blanchetii*	(Barbosa et al., 2015)
Vitaceae	*Cissus sicyoides*	(Beltrame et al., 2002)
	Vitis vinifera	(Ribeiro et al., 2015)
	Vitis labrusca	
Xanthorrhoeaceae	*Asphodeline anatolica*	(Marino et al., 2016)
Zingiberaceae	*Alpinia officinarum*	(Zou et al., 2016)
	Cautleya spicata	(Semwal et al., 2015)
	Etlingera elatior	(Ghasemzadeh et al., 2015b)
	Naudea officinalis	(Xie et al., 2011)
Zygophyllaceae	*Faconia arabica*	(El-Wakil, 2007)
	Faconia taeckholmiana	(Huang et al., 2010; Ibrahim et al., 2008)

3. Biological Activities of Kaempferol

Several epidemiological studies reported the possible association between the consumption of food containing kaempferol and a reduced risk of developing several disorders, including lung cancer (Cui et al., 2008), gastric cancer (Garcia-Closas et al., 1999), pancreatic cancer (Nöthlings et al., 2007), ovarian cancer (Gates et al., 2007) and cardiovascular diseases (Geleijnse et al., 2002; Marniemi et al., 2005). In a previous case study (Cui et al., 2008) observed that the consumption of kaempferol rich foods (approximately 2 mg kaempferol/day) was inversely associated with lung cancer risk.

Numerous reports both *in vitro* and *in vivo* have shown that kaempferol, some glycosides of kaempferol, and several kaempferol-containing plants have antioxidant activity (Aniya et al., 2005; Bonina et al., 2002; Kampkötter et al., 2007) and anti-inflammatory activity (De Melo et al., 2005; Hossen et al., 2015a; Hossen et al., 2015b; Hossen et al., 2015c; Hossen et al., 2015d; Kim et al., 2010; Küpeli et al., 2007; Orhan et al., 2007; Rao et al., 2008). For instance, studies with methanolic extract from the aerial parts of kaempferol-containing plants *Phyllanthus acidus* (Hossen et al., 2015b; Hossen et al., 2015c) and *Persicaria chinensis* (Hossen et al., 2015a; Hossen et al., 2015d) exhibited strong anti-gastritis and hepatoprotective activity as well as *in vivo* and *in vitro* anti-inflammatory activity by targeting the Src/Syk of NFκB. Evidence suggests that some kaempferol glycosides and several kaempferol-containing plants have antidiabetic activity (Anwar et al., 2007) and may prevent diabetic complications (Ghaffari and Mojab, 2007). For example, kaempferitrin and kaempferol 3-neohesperidoside isolated from *Cyathea phalerata* stems, showed a significant hypoglycemic effect in diabetic rats (Cazarolli et al., 2006) which may be mediated by stimulation of glycogen synthesis.

In addition to having cancer chemopreventive properties, kaempferol has shown activity which may relevant as a cancer therapy. Kaempferol and some glycosides of kaempferol induces cancer cell death in different tissues, including lung (Conforti et al., 2009; Leung et al., 2007), breast (Kang et al., 2009; Kim et al., 2008), colon (Li et al., 2009b), prostate (Brusselmans et al., 2005), liver (Mylonis et al., 2010), pancreas (Zhang et al., 2008),blood/lymph (Benyahia et al., 2004; Li et al., 2007; Wang et al., 2006), skin (Conforti et al., 2009; Li et al., 2007), brain (Jeong et al., 2009b), uterus (Li et al., 2007), and ovary (Luo et al., 2010).

4. Bioavailability

Voluminous preclinical research confirmed that kaempferol showed holistic pharmacological activities including antioxidant, anti-inflammatory, antimicrobial, anticancer, cardioprotective, neuroprotective, antidiabetic, anti-osteoporotic, estrogenic/antiestrogenic, anxiolytic, analgesic and antiallergic activities. This flavonoid is reported to have admirable antioxidative activity and can react with H_2O_2, HOCL, superoxide and nitric oxide. Molecular investigations have shown that kaempferol modulates a number of key elements in cellular signal transduction pathways linked to apoptosis, angiogenesis, inflammation, and metastasis and shows lower toxicity in comparison to standard chemotherapy drugs. The novel anti-inflammatory activities of this compound were exhibited by inhibiting cell function as well as proinflammatory cytokines and chemokine expression, and by directly targeting kinases. Kaempferol demonstrated poor oral bioavailability and is commonly metabolized into the forms of methyl, sulfate or glucuronide but in combination with other anticancer agents, it can boost its anticancer affinity. For instance, the combination of kaempferol with quercetin significantly enhances the anticancer activities of quercetin by blocking the efflux of quercetin, signifying that kaempferol increases the bioavailability of different anticancer drugs.

Conclusion

Most research was conducted using doses far beyond the oral bioavailability of kaempferol. Therefore, it seems very difficult to draw a conclusion about the most effective doses of this flavonoid. Further studies should focus on effective doses of kaempferol for clinical trials and aim to solve the problems related to low bioavailability, permeability and safe dosage requirement to offer this flavonoid as a prospective novel candidate for future drug development.

References

Aaby K, Ekeberg D and Skrede G (2007) Characterization of phenolic compounds in strawberry (Fragaria× ananassa) fruits by different HPLC detectors and contribution of individual compounds to total antioxidant capacity. *Journal of Agricultural and Food Chemistry* 55:4395-4406.

Abd El-Latif RR, Mansour RM, Sharaf M and Farag A (2014) Three new flavonol glycosides from Suaeda maritima. *Journal of Asian natural products research* 16:434-439.

Abdelfattah MS, Toume K, Ahmed F, Sadhu SK and Ishibashi M (2010) Cucullamide, a new putrescine bisamide from Amoora cucullata. *Chemical and Pharmaceutical Bulletin* 58:1116-1118.

Abe F, Iwase Y, Yamauchi T, Yahara S and Nohara T (1995) Flavonol sinapoyl glycosides from leaves of Thevetia peruviana. *Phytochemistry* 40:577-581.

Abreu PM, Matthew S, González T, Vanickova L, Costa D, Gomes A, Segundo MA and Fernandes E (2008) Isolation and identification of antioxidants from Pedilanthus tithymaloides. *Journal of natural medicines* 62:67-70.

Adesina S, Idowu O, Ogundaini A, Oladimeji H, Olugbade T, Onawunmi G and Pais M (2000) Antimicrobial constituents of the leaves of Acalypha wilkesiana and Acalypha hispida. *Phytotherapy Research* 14:371-374.

Ahmed KM, Marzouk M, el-Khrisy E, Wahab SA and El-Din S (2000) A new flavone diglycoside from Carthamus tinctorius seeds. *Die Pharmazie* 55:621-622.

Akak CM, Djama CM, Nkengfack AE, Tu P-F and Lei L-D (2010) New coumarin glycosides from the leaves of Diospyros crassiflora (Hiern). *Fitoterapia* 81:873-877.

Al-Oqail M, Hassan WH, Ahmad MS and Al-Rehaily AJ (2012) Phytochemical and biological studies of Solanum schimperianum Hochst. *Saudi Pharmaceutical Journal* 20:371-379.

Al-Taweel AM, Abdel-Kader MS, Fawzy GA, Perveen S, Maher HM, Al-Zoman NZ, Al-Shehri MM, Al-Johar H and Al-Showiman H (2015) Isolation of flavonoids from Delonix elata and determination of its rutin content using capillary electrophoresis. *Pakistan journal of pharmaceutical sciences* 28:1897-1903.

Al-Taweel AM, Fawzy GA and Perveen S (2012) Cytotoxic Flavonoid Glycosides from Rapistrum rugosum L. *Iranian journal of pharmaceutical research: IJPR* 11:839.

Ali A, Mohamed M, Kamel M, Fouad M and Spring O (1998) Studies on Securigera securidacea (L.) Deg. et Dorfl.(Fabaceae) seeds, an antidiabetic Egyptian folk medicine. *Die Pharmazie* 53:710-715.

Amakura Y, Yoshimura M, Sugimoto N, Yamazaki T and Yoshida T (2009) Marker constituents of the natural antioxidant Eucalyptus leaf extract for the evaluation of food additives. *Bioscience, biotechnology, and biochemistry* 73:1060-1065.

Amaral JS, Ferreres F, Andrade PB, Valentão P, Pinheiro C, Santos A and Seabra R (2005) Phenolic profile of hazelnut (Corylus avellana L.) leaves cultivars grown in Portugal. *Natural product research* 19:157-163.

Aniya Y, Koyama T, Miyagi C, Miyahira M, Inomata C, Kinoshita S and Ichiba T (2005) Free radical scavenging and hepatoprotective actions of the medicinal herb, Crassocephalum crepidioides from the Okinawa Islands. *Biological and Pharmaceutical Bulletin* 28:19-23.

Anwar F, Latif S, Ashraf M and Gilani AH (2007) Moringa oleifera: a food plant with multiple medicinal uses. *Phytotherapy research* 21:17-25.

Arriaga A, MESQUITA AC, Pouliquen Y, LIMA RA, Cavalcante SH, CARVALHO MG, SIQUEIRA JA, Alegrio LV and Braz-Filho R (2002) Chemical constituents of Simarouba versicolor. *Anais da Academia Brasileira de Ciências* 74:415-424.

Aung H, Chia L, Goh N, Chia T, Ahmed A, Pare P and Mabry T (2002) Phenolic constituents from the leaves of the carnivorous plant Nepenthes gracilis. *Fitoterapia* 73:445-447.

Autore G, Rastrelli L, Lauro MR, Marzocco S, Sorrentino R, Sorrentino U, Pinto A and Aquino R (2001) Inhibition of nitric oxide synthase expression by a methanolic extract of Crescentia alata and its derived flavonols. *Life sciences* 70:523-534.

Badria FA, Ameen M and Akl MR (2007) Evaluation of cytotoxic compounds from Calligonum comosum L. growing in Egypt. *Zeitschrift für Naturforschung C* 62:656-660.

Bai N, He K, Roller M, Lai C-S, Shao X, Pan M-H and Ho C-T (2010) Flavonoids and phenolic compounds from Rosmarinus officinalis. *Journal of agricultural and food chemistry* 58:5363-5367.

Barbosa JS, Almeida VM, Marçal RM and Branco A (2015) RP-HPLC-DAD-MS (n) Analysis and Butyrylcholinesterase Inhibitory Activity of Barbacenia blanchetii Extracts. *Natural product communications* 10:983-986.

Barile E, Capasso R, Izzo AA, Lanzotti V, Sajjadi SE and Zolfaghari B (2005) Structure-activity relationships for saponins from Allium hirtifolium and Allium elburzense and their antispasmodic activity. *Planta medica* 71:1010-1018.

Behbahani SS, Pourazar A and M Shanehsazzadeh M (2014) In vitro anti-HIV-1 activities of kaempferol and kaempferol-7-O-glucoside isolated from Securigera securidaca. *Research in Pharmaceutical Sciences* 9:463-469.

Beltrame F, Ferreira A and Cortez D (2002) Coumarin glycoside from Cissus sicyoides. *Natural product letters* 16:213-216.

Bennett RN, Rosa EA, Mellon FA and Kroon PA (2006) Ontogenic profiling of glucosinolates, flavonoids, and other secondary metabolites in Eruca sativa (salad rocket), Diplotaxis erucoides (wall rocket), Diplotaxis tenuifolia (wild rocket), and Bunias orientalis (Turkish rocket). *Journal of agricultural and food chemistry* 54:4005-4015.

Benyahia S, Benayache S, Benayache F, Quintana J, López M, León F, Hernández JC, Estévez F and Bermejo J (2004) Isolation from Eucalyptus o ccidentalis and Identification of a New Kaempferol Derivative that Induces Apoptosis in Human Myeloid Leukemia Cells. *Journal of natural products* 67:527-531.

Berhow MA, Bennett RD, Poling SM, Vannier S, Hidaka T and Omura M (1994) Acylated flavonoids in callus cultures of Citrus aurantifolia. *Phytochemistry* 36:1225-1227.

Bonina F, Puglia C, Ventura D, Aquino R, Tortora S, Sacchi A, Saija A, Tomaino A, Pellegrino M and de Capariis P (2002) In vitro antioxidant and in vivo photoprotective effects of a lyophilized extract of Capparis spinosa L. buds. *Journal of cosmetic science* 53:321-336.

Brusselmans K, Vrolix R, Verhoeven G and Swinnen JV (2005) Induction of cancer cell apoptosis by flavonoids is associated with their ability to inhibit fatty acid synthase activity. *Journal of Biological Chemistry* 280:5636-5645.

Calvo TR, Cardoso CRP, da Silva Moura AC, dos Santos LC, Colus IMS, Vilegas W and Varanda EA (2011) Mutagenic activity of Indigofera truxillensis and I. suffruticosa aerial parts. *Evidence-Based Complementary and Alternative Medicine* 2011.

Calzada F (2005) Additional antiprotozoal constituents from Cuphea pinetorum, a plant used in Mayan traditional medicine to treat diarrhoea. *Phytotherapy Research* 19:725-727.

Calzada F, Cedillo-Rivera R and Mata R (2001) Antiprotozoal Activity of the Constituents of Conyza f ilaginoides 1. *Journal of natural products* 64:671-673.

Carotenuto A, De Feo V, Fattorusso E, Lanzotti V, Magno S and Cicala C (1996) The flavonoids of Allium ursinum. *Phytochemistry* 41:531-536.

Carotenuto A, Fattorusso E, Lanzotti V, Magno S, De Feo V and Cicala C (1997) The flavonoids of Allium neapolitanum. *Phytochemistry* 44:949-957.

Cazarolli LH, Zanatta L, Jorge AP, de Sousa E, Horst H, Woehl VM, Pizzolatti MG, Szpoganicz B and Silva FRMB (2006) Follow-up studies on glycosylated flavonoids and their complexes with vanadium: their anti-hyperglycemic potential role in diabetes. *Chemico-biological interactions* 163:177-191.

Cetto AA, Wiedenfeld H, Revilla MC and Sergio IA (2000) Hypoglycemic effect of Equisetum myriochaetum aerial parts on streptozotocin diabetic rats. *Journal of Ethnopharmacology* 72:129-133.

Champavier Y, Allais DP, Chulia AJ and Kaouadji M (2000) Acetylated and non-acetylated flavonol triglycosides from Galega officinalis. *CHEMICAL AND PHARMACEUTICAL BULLETIN-TOKYO-* 48:281-282.

Chang C-C and Lee S-S (2015) Peracylated Glucosyl Kaempferols from Pasania dodonfifolia Leaf. *Natural product communications* 10:1373-1375.

Chang F-R, Wei J-L, Teng C-M and Wu Y-C (1998) Antiplatelet Aggregation Constituents from Annona p urpurea. *Journal of natural products* 61:1457-1461.

Chatzopoulou A, Karioti A, Gousiadou C, Lax Vivancos V, Kyriazopoulos P, Golegou S and Skaltsa H (2010) Depsides and other polar constituents from Origanum dictamnus L. and their in vitro antimicrobial activity in clinical strains. *Journal of agricultural and food chemistry* 58:6064-6068.

Chen G, Lu H, Wang C, Yamashita K, Manabe M, Meng Z, Xu S and Kodama H (2002) Effect of five flavonoid compounds isolated from leaves of Diospyros kaki on stimulus-induced superoxide generation and tyrosyl phosphorylation of proteins in human neutrophils. *Clinica chimica acta* 326:169-175.

Chen GY and Nunez G (2010) Sterile inflammation: sensing and reacting to damage. *Nat Rev Immunol* 10:826-837.

Chen Y-J, Xie G-Y, Xu G-K, Dai Y-Q, Shi L and Qin M-J (2015) Chemical Constituents of Pyrrosia calvata. *Natural product communications* 10:1191-1193.

Cheng M, Meng L, Zhou X, Zou H, Yu S and Zhou G (2014) [Chemical constituents of flavonoids and their glycosides in Melastoma dodecandrum]. *Zhongguo Zhong yao za zhi= Zhongguo zhongyao zazhi= China journal of Chinese materia medica* 39:3301-3305.

Chien M, Svoboda G, Schiff P, Slatkin D and Knapp J (1979) Chemical constituents of Echites hirsuta (Apocynaceae). *Journal of pharmaceutical sciences* 68:247-249.

Choi C-W, Jung HA, Kang SS and Choi JS (2007) Antioxidant constituents and a new triterpenoid glycoside fromFlos Lonicerae. *Archives of pharmacal research* 30:1-7.

Choi SZ, Choi SU and Lee KR (2004) Pytochemical constituents of the aerial parts fromsolidago virga-aurea var. gigantea. *Archives of pharmacal research* 27:164-168.

Christopoulou C, Graikou K and Chinou I (2008) Chemosystematic value of chemical constituents from Scabiosa hymettia (Dipsacaceae). *Chemistry & biodiversity* 5:318-323.

Chu H, He W, Zhang Y, Ji C and Tan N (2011) [Flavonoids and nor-sesquiterpenes of Pedicularis densispica]. *Zhongguo Zhong yao za zhi= Zhongguo zhongyao zazhi= China journal of Chinese materia medica* 36:2672-2675.

Clericuzio M, Tinello S, Burlando B, Ranzato E, Martinotti S, Cornara L and La Rocca A (2012) Flavonoid oligoglycosides from Ophioglossum vulgatum L. having wound healing properties. *Planta Medica-Natural Products and MedicinalPlant Research* 78:1639.

Conforti F, Menichini F, Rigano D and Senatore F (2009) Antiproliferative activity on human cancer cell lines after treatment with polyphenolic compounds isolated from Iris pseudopumila flowers and rhizomes. *Zeitschrift für Naturforschung C* 64:490-494.

Cordell G, Lyon R, Fong H, Benoit P and Farnsworth N (1976) Biological and phytochemical investigations of Dianthus barbatus cv. "China Doll" (Caryophyllaceae). *Lloydia* 40:361-363.

Corea G, Fattorusso E and Lanzotti V (2003) Saponins and Flavonoids of Allium t riquetrum. *Journal of natural products* 66:1405-1411.

Crublet M-L, Long C, Sévenet T, Hadi HA and Lavaud C (2003) Acylated flavonol glycosides from leaves of Planchonia grandis. *Phytochemistry* 64:589-594.

Cuendet M, Potterat O and Hostettmann K (2001) Flavonoids and phenylpropanoid derivatives from Campanula barbata. *Phytochemistry* 56:631-636.

Cui Y, Morgenstern H, Greenland S, Tashkin DP, Mao JT, Cai L, Cozen W, Mack TM, Lu QY and Zhang ZF (2008) Dietary flavonoid intake and lung cancer—A population-based case-control study. *Cancer* 112:2241-2248.

Dai L-M, Zhao C-C, Jin H-z, Tang J, Shen Y-H, Li H-L, Peng C-Y and Zhang W-D (2008) A new ferulic acid ester and other constituents from Dracocephalum peregrinum. *Archives of pharmacal research* 31:1325-1329.

Dall'Acqua S, Tomè F, Vitalini S, Agradi E and Innocenti G (2009) In vitro estrogenic activity of Asplenium trichomanes L. extracts and isolated compounds. *Journal of ethnopharmacology* 122:424-429.

Das S and Ganapaty S (2015) Phytochemical evaluation of roots of polygonum viscosum buch-ham. *Indian journal of pharmaceutical sciences* 77:352.

Dat NT, Dang NH and Thanh LN (2016) New flavonoid and pentacyclic triterpene from Sesamum indicum leaves. *Natural product research* 30:311-315.

de Carvalho MG, Carvalho Cranchi D, Kingston DG and Werle AA (2001) Proposed active constituents of Dipladenia martiana. *Phytotherapy Research* 15:715-717.

De Laurentis N, Stefanizzi L, Milillo M and Tantillo G (1997) Flavonoids from leaves of Olea europaea L. cultivars, in *Annales pharmaceutiques françaises* pp 268-273.

De Leo M, Braca A, De Tommasi N, Norscia I, Morelli I, Battinelli L and Mazzanti G (2004) Phenolic compounds from Baseonema acuminatum leaves: isolation and antimicrobial activity. *Planta medica* 70:841-846.

De Melo GO, Malvar DdC, Vanderlinde FA, Pires PA, Côrtes WS, Germano Filho P, Muzitano MF, Kaiser CR and Costa SS (2005) Phytochemical and pharmacological study of Sedum dendroideum leaf juice. *Journal of ethnopharmacology* 102:217-220.

De Simone F, Dini A, Pizza C, Saturnino P and Schettino O (1990) Two flavonol glycosides from Chenopodium quinoa. *Phytochemistry* 29:3690-3692.

Deng J-Z, Marshall R, Jones SH, Johnson RK and Hecht SM (2002) DNA-damaging agents from Crypteronia paniculata. *Journal of natural products* 65:1930-1932.

Deng S, Palu AK, West BJ, Su CX, Zhou B-N and Jensen JC (2007) Lipoxygenase inhibitory constituents of the fruits of noni (Morinda citrifolia) collected in Tahiti. *Journal of natural products* 70:859-862.

Diaz JG and Herz W (2010) Acylated flavonol tetraglycosides from Delphinium gracile. *Phytochemistry* 71:463-468.

Dini A, Rastrelli L, Saturnino P and Schettino O (1991) Minor components in food plants--Note I. Flavonol glycosides from Ullucus tuberosus. *Bollettino della Società italiana di biologia sperimentale* 67:1053.

Djouossi MG, Ngnokam D, Kuiate J-R, Tapondjou LA, Harakat D and Voutquenne-Nazabadioko L (2015) Antimicrobial and antioxidant flavonoids from the leaves of Oncoba spinosa Forssk.(Salicaceae). *BMC complementary and alternative medicine* 15:134.

Drewes SE and Taylor CW (1994) Methylated A-type proanthocyanidins and related metabolites from Cassipourea gummiflua. *Phytochemistry* 37:551-555.

Du X-M, Sun N-Y and Shoyama Y (2000) Flavonoids from Goodyera schlechtendaliana. *Phytochemistry* 53:997-1000.

Duan S, Tang S, Qin N and Duan H (2012) [Chemical constituents of Phymatopteris hastate and their antioxidant activity]. *Zhongguo Zhong yao za zhi= Zhongguo zhongyao zazhi= China journal of Chinese materia medica* 37:1402-1407.

Dudek-Makuch M and Matławska I (2011) Flavonoids from the flowers of Aesculus hippocastanum. *Acta Pol Pharm* 68:403-408.

Dumlu M, Gurkan E and Tuzlaci E (2008) Chemical composition and antioxidant activity of Campanula alliariifolia. *Natural product research* 22:477-482.

DuPont MS, Mondin Z, Williamson G and Price KR (2000) Effect of variety, processing, and storage on the flavonoid glycoside content and composition of lettuce and endive. *Journal of agricultural and food chemistry* 48:3957-3964.

El-Sayed NH, Wojcińska M, Drost-Karbowska K, Matławska I, Williams J and Mabry TJ (2002) Kaempferol triosides from Silphium perfoliatum. *Phytochemistry* 60:835-838.

El-Wakil EA (2007) Phytochemical and molluscicidal investigations of Fagonia arabica. *Zeitschrift für Naturforschung C* 62:661-667.

El Dib R, Soliman H, Hussein M and Attia H (2015) Two New Flavonoids and Biological Activity of Astragalus abyssinicus (Hochst.) Steud. ex A. Rich. Aerial Parts. *Drug research* 65:259-265.

Ezzat SM, Choucry MA and Kandil ZA (2015) Antibacterial, antioxidant, and topical anti-inflammatory activities of Bergia ammannioides: A wound-healing plant. *Pharmaceutical biology*:1-10.

Fang S-H, Rao YK and Tzeng Y-M (2005) Inhibitory effects of flavonol glycosides from Cinnamomum osmophloeum on inflammatory mediators in LPS/IFN-γ-activated murine macrophages. *Bioorganic & medicinal chemistry* 13:2381-2388.

Fang X-K, Gao J and Zhu D-N (2008) Kaempferol and quercetin isolated from Euonymus alatus improve glucose uptake of 3T3-L1 cells without adipogenesis activity. *Life sciences* 82:615-622.

Fatima T, Kesari V, Watt I, Wishart D, Todd JF, Schroeder WR, Paliyath G and Krishna P (2015) Metabolite profiling and expression analysis of flavonoid, vitamin C and tocopherol biosynthesis genes in the antioxidant-rich sea buckthorn (Hippophae rhamnoides L.). *Phytochemistry* 118:181-191.

Fattorusso E, Lanzotti V, Taglialatela-Scafati O and Cicala C (2001) The flavonoids of leek, Allium porrum. *Phytochemistry* 57:565-569.

Feng W-S, Zhang Q-B, Zheng X-K, Chen H, Zhang Y-L and Zhang C-L (2012) A new acylated flavonol glycoside from the aerial parts of Cardamine tangutorum. *Journal of Asian natural products research* 14:805-810.

Ferrari J, Terreaux C, Sahpaz S, Msonthi JD, Wolfender J-L and Hostettmann K (2000) Benzophenone glycosides from Gnidia involucrata. *Phytochemistry* 54:883-889.

Ferreira RO, da Silva TMS and de Carvalho MG (2015) New Polyprenylated Phloroglucinol and Other Compounds Isolated from the Fruits of Clusia nemorosa (Clusiaceae). *Molecules* 20:14326-14333.

Ferreres F, Taveira M, Pereira DM, Valentão Pc and B. Andrade P (2010) Tomato (Lycopersicon esculentum) seeds: new flavonols and cytotoxic effect. *Journal of agricultural and food chemistry* 58:2854-2861.

Filomeni G, Graziani I, De Zio D, Dini L, Centonze D, Rotilio G and Ciriolo MR (2012) Neuroprotection of kaempferol by autophagy in models of rotenone-mediated acute toxicity: possible implications for Parkinson's disease. *Neurobiology of aging* 33:767-785.

Francescato LN, Debenedetti SL, Schwanz TG, Bassani VL and Henriques AT (2013) Identification of phenolic compounds in Equisetum giganteum by LC–ESI-MS/MS and a new approach to total flavonoid quantification. *Talanta* 105:192-203.

Francis JA, Rumbeiha W and Nair MG (2004) Constituents in Easter lily flowers with medicinal activity. *Life sciences* 76:671-683.

Fu M, Deng D, Huang R, Zhang N, Su Z and Qiu SX (2013) A new flavanocoumarin from the root of Flemingia philippinensis. *Natural product research* 27:1237-1241.

Fu Y, Huang L, Wang X, Li X, Li K, Wu S and Liu Y (2015) [Studies on chemical constituents from Psychotria straminea]. *Zhongguo Zhong yao za zhi= Zhongguo zhongyao zazhi= China journal of Chinese materia medica* 40:2138-2143.

Gamez EJ, Luyengi L, Lee SK, Zhu L-F, Zhou B-N, Fong HH, Pezzuto JM and Kinghorn AD (1998) Antioxidant Flavonoid Glycosides from Daphniphyllum calycinum 1. *Journal of Natural products* 61:706-708.

Gao H, Kuroyanagi M, Wu L, Kawahara N, Yasuno T and Nakamura Y (2002) Antitumor-promoting constituents from Dioscorea bulbifera L. in JB6 mouse epidermal cells. *Biological and Pharmaceutical Bulletin* 25:1241-1243.

Gao L, Tian H, Lv P, Wang J and Wang Y (2015) [Chemical constituents of Sapium sebiferum leaves]. *Zhongguo Zhong yao za zhi= Zhongguo zhongyao zazhi= China journal of Chinese materia medica* 40:1518-1522.

Garcia-Closas R, Gonzalez CA, Agudo A and Riboli E (1999) Intake of specific carotenoids and flavonoids and the risk of gastric cancer in Spain. *Cancer Causes & Control* 10:71-75.

Gates MA, Tworoger SS, Hecht JL, De Vivo I, Rosner B and Hankinson SE (2007) A prospective study of dietary flavonoid intake and incidence of epithelial ovarian cancer. *International Journal of Cancer* 121:2225-2232.

Gayosso-De-Lucio JA, Torres-Valencia JM, Cerda-Garcia-Rojas CM and Joseph-Nathan P (2010) Ellagitannins from Geranium potentillaefolium and G. bellum. *Natural product communications* 5:531-534.

Geleijnse JM, Launer LJ, van der Kuip DA, Hofman A and Witteman JC (2002) Inverse association of tea and flavonoid intakes with incident myocardial infarction: the Rotterdam Study. *The American journal of clinical nutrition* 75:880-886.

Ghaffari MA and Mojab S (2007) Influence of flavonols as in vitro on low density lipoprotein glycation. *Iranian Biomedical Journal* 11:185-191.

Ghasemzadeh A, Jaafar HZ and Rahmat A (2015a) Phytochemical constituents and biological activities of different extracts of Strobilanthes crispus (L.) Bremek leaves grown in different locations of Malaysia. *BMC complementary and alternative medicine* 15:422.

Ghasemzadeh A, Jaafar HZ, Rahmat A and Ashkani S (2015b) Secondary metabolites constituents and antioxidant, anticancer and antibacterial activities of Etlingera elatior (Jack) RM Sm grown in different locations of Malaysia. *BMC complementary and alternative medicine* 15:1.

Goel R, Pandey V, Dwivedi S and Rao Y (1988) Antiinflammatory and antiulcer effects of kaempferol, a flavone, isolated from Rhamnus procumbens. *Indian journal of experimental biology* 26:121-124.

Gogna N, Hamid N and Dorai K (2015) Metabolomic profiling of the phytomedicinal constituents of Carica papaya L. leaves and seeds by 1 H

NMR spectroscopy and multivariate statistical analysis. *Journal of pharmaceutical and biomedical analysis* 115:74-85.

Gohar AA, Maatooq GT and Niwa M (2000) Two flavonoid glycosides from Chenopodium murale. *Phytochemistry* 53:299-303.

Grančai D, Fialová S, Huszár V, Karlovská J and Svajdlenka E (2014) [Flavonoids-main constituents of the leaves of Philadelphus tenuifolius Rupr. et Maxim]. *Ceska a Slovenska farmacie: casopis Ceske farmaceuticke spolecnosti a Slovenske farmaceuticke spolecnosti* 63:120-122.

Grundmann O, Nakajima J-I, Kamata K, Seo S and Butterweck V (2009) Kaempferol from the leaves of Apocynum venetum possesses anxiolytic activities in the elevated plus maze test in mice. *Phytomedicine* 16:295-302.

Gurbuz I, Yesilada E and Ito S (2009) An anti-ulcerogenic flavonol diglucoside from Equisetum palustre L. *Journal of ethnopharmacology* 121:360-365.

Habbu P, Mahadevan K, Shastry R and Manjunatha H (2009) Antimicrobial activity of flavanoid sulphates and other fractions of Argyreia speciosa (Burm. f) Boj.

Harbaum B, Hubbermann EM, Wolff C, Herges R, Zhu Z and Schwarz K (2007) Identification of Flavonoids and Hydroxycinnamic Acids in Pak Choi Varieties (Brassica campestris L. ssp. chinensis var. communis) by HPLC–ESI-MS n and NMR and Their Quantification by HPLC–DAD. *J of Agricultural and Food Chemistry* 55:8251-8260.

Hashiguchi T, Sakakibara Y, Shimohira T, Kurogi K, Yamasaki M, Nishiyama K, Akashi R, Liu M-C and Suiko M (2013) Identification of a novel flavonoid glycoside sulfotransferase in Arabidopsis thaliana. *Journal of biochemistry*:mvt102.

Hashimoto N, Ohsawa R, Kitajima J and Iwashina T (2015) New flavonol glycosides from the leaves and flowers of Primula sieboidii. *Natural product communications* 10:421-423.

Hassan RA, Tawfik WA and Abou-Setta LM (2013) The flavonoid constituents of Leucaena Leucocephala growning in Egypt, and their biological activity. *African Journal of Traditional, Complementary and Alternative Medicines* 11:67-72.

Hatano T, Yasuhara T, Fukuda T, Noro T and Okuda T (1989) Phenolic constituents of licorice. II. Structures of licopyranocoumarin, licoarylcoumarin and glisoflavone, and inhibitory effects of licorice phenolics on xanthine oxidase. *Chemical & pharmaceutical bulletin* 37:3005-3009.

Ho HM, Chen R, Huang Y and Chen ZY (2002) Vascular effects of a soy leaves (Glycine max) extract and kaempferol glycosides in isolated rat carotid arteries. *Planta medica* 68:487-491.

Hossen MJ, Baek KS, Kim E, Yang WS, Jeong D, Kim JH, Kweon DH, Yoon DH, Kim TW, Kim JH and Cho JY (2015a) In vivo and in vitro anti-inflammatory activities of Persicaria chinensis methanolic extract targeting Src/Syk/NF-kappaB. *J Ethnopharmacol* 159:9-16.

Hossen MJ, Jeon SH, Kim SC, Kim JH, Jeong D, Sung NY, Yang S, Baek KS, Kim JH, Yoon DH, Song WO, Yoon KD, Cho SH, Lee S, Kim JH and Cho JY (2015b) In vitro and in vivo anti-inflammatory activity of Phyllanthus acidus methanolic extract. *J Ethnopharmacol* 168:217-228.

Hossen MJ, Kim M-Y, Kim J-H and Cho JY (2015c) AP-1-Targeted Inhibition of Macrophage Function and Lipopolysaccharide/D-galactosamine-Induced Hepatitis by Phyllanthus acidus Methanolic Extract. *The American journal of Chinese medicine*:1-22.

Hossen MJ, Kim SC, Son YJ, Baek KS, Kim E, Yang WS, Jeong D, Park JG, Kim HG, Chung WJ, Yoon K, Ryou C, Lee SY, Kim JH and Cho JY (2015d) AP-1-Targeting Anti-Inflammatory Activity of the Methanolic Extract of Persicaria chinensis. *Evidence-based complementary and alternative medicine: eCAM* 2015:608126.

Hu J, Zhang L and Zhao Q (2011) [Chemical constituents of Excoecaria acerifclia]. *Zhongguo Zhong yao za zhi= Zhongguo zhongyao zazhi= China journal of Chinese materia medica* 36:1969-1974.

Hu M, Ning Y, Wu A, Xia Y, Zhu C and Lin C (2013) [Study on chemical constituents of Callicarpa peii]. *Zhong yao cai= Zhongyaocai= Journal of Chinese medicinal materials* 36:563-566.

Huang H-C, Syu K-Y and Lin J-K (2010) Chemical composition of Solanum nigrum linn extract and induction of autophagy by leaf water extract and its major flavonoids in AU565 breast cancer cells. *Journal of agricultural and food chemistry* 58:8699-8708.

Hung TM, Dang NH, Kim JC, Choi JS, Lee HK and Min B-S (2009) Phenolic glycosides from Alangium salviifolium leaves with inhibitory activity on LPS-induced NO, PGE 2, and TNF-α production. *Bioorganic & Medicinal Chemistry Letters* 19:4389-4393.

Hussein S (2004) Flavonoid and methoxyellagic acid sodium sulphates from Frankenia laevis L. *Die Pharmazie-An International Journal of Pharmaceutical Sciences* 59:484-487.

Hwang E-I, Ahn B-T, Lee H-B, Kim Y-K, Lee K-S, Bok S-H, Kim Y-T and Kim S-U (2001) Inhibitory activity for chitin synthase II from Saccharomyces cerevisiae by tannins and related compounds. *Planta medica* 67:501-504.

Ibrahim LF, Kawashty SA, El-Hagrassy AM, Nassar MI and Mabry TJ (2008) A new kaempferol triglycoside from Fagonia taeckholmiana: cytotoxic activity of its extracts. *Carbohydrate research* 343:155-158.

Ikechukwu O and Ifeanyi O (2015) The Antidiabetic Effects of The Bioactive Flavonoid (Kaempferol-3-O-β-D-6 {P-Coumaroyl} Glucopyranoside) Isolated From Allium cepa. *Recent patents on anti-infective drug discovery.*

Iwashina T, Tamura MN, Murai Y and Kitajima J (2013) New flavonol glycosides from the leaves of Triantha japonica and Tofieldia nuda. *Natural product communications* 8:1251-1254.

Je Ma C, Jung WJ, Lee KY, Kim YC and Sung SH (2009) Calpain inhibitory flavonoids isolated from Orostachys japonicus. *Journal of enzyme inhibition and medicinal chemistry* 24:676-679.

Jeong HJ, Ryu YB, Park S-J, Kim JH, Kwon H-J, Kim JH, Park KH, Rho M-C and Lee WS (2009a) Neuraminidase inhibitory activities of flavonols isolated from Rhodiola rosea roots and their in vitro anti-influenza viral activities. *Bioorganic & medicinal chemistry* 17:6816-6823.

Jeong JC, Kim MS, Kim TH and Kim YK (2009b) Kaempferol induces cell death through ERK and Akt-dependent down-regulation of XIAP and survivin in human glioma cells. *Neurochemical research* 34:991-1001.

Jia L, Jing L-L, Zhou S-A and Kong D-Y (2011) Three new flavonoid glycosides from Urena lobata. *Journal of Asian natural products research* 13:907-914.

Jia L, Wang J, Lv C, Xu T, He L, Dong Y and Lu J (2013) Two new compounds from Potentilla multicaulis Bunge. *Natural product research* 27:1361-1365.

Jiang H, Zhan W, Liu X and Jiang S (2008) Antioxidant activities of extracts and flavonoid compounds from Oxytropis falcate Bunge. *Natural product research* 22:1650-1656.

Jiang R-W, Zhou J-R, Hon P-M, Li S-L, Zhou Y, Li L-L, Ye W-C, Xu H-X, Shaw P-C and But PP-H (2007) Lignans from Dysosma versipellis with Inhibitory Effects on Prostate Cancer Cell Lines⊥. *Journal of natural products* 70:283-286.

Kaewamatawong R, Kitajima M, Kogure N and Takayama H (2008) Flavonols from Bauhinia malabarica. *Journal of natural medicines* 62:364-365.

Kamel M, Ohtani K, Hasanain H, Mohamed M, Kasai R and Yamasaki K (2000) Monoterpene and pregnane glucosides from Solenostemma argel. *Phytochemistry* 53:937-940.

Kampkötter A, Nkwonkam CG, Zurawski RF, Timpel C, Chovolou Y, Wätjen W and Kahl R (2007) Effects of the flavonoids kaempferol and fisetin on thermotolerance, oxidative stress and FoxO transcription factor DAF-16 in the model organism Caenorhabditis elegans. *Archives of toxicology* 81:849-858.

Kang G-Y, Lee E-R, Kim J-H, Jung JW, Lim J, Kim SK, Cho S-G and Kim KP (2009) Downregulation of PLK-1 expression in kaempferol-induced apoptosis of MCF-7 cells. *European journal of pharmacology* 611:17-21.

Kaouadji M, Doucouré A, Mariotte A-M, Chulia AJ and Thomasson F (1990) Flavonol triglycosides from Blackstonia peroliata. *Phytochemistry* 29:1283-1286.

Karamenderes C, Bedir E, Pawar R, Baykan S and Khan IA (2007) Elemanolide sesquiterpenes and eudesmane sesquiterpene glycosides from Centaurea hierapolitana. *Phytochemistry* 68:609-615.

Kassem ME, Afifi MS, Marzouk MM and Mostafa MA (2013) Two new flavonol glycosides and biological activities of Diplotaxis harra (Forssk.) Boiss. *Natural product research* 27:2272-2280.

Kataoka M, Hirata K, Kunikata T, Ushio S, Iwaki K, Ohashi K, Ikeda M and Kurimoto M (2001) Antibacterial action of tryptanthrin and kaempferol, isolated from the indigo plant (Polygonum tinctorium Lour.), against Helicobacter pylori-infected Mongolian gerbils. *Journal of gastroenterology* 36:5-9.

Kazuma K, Noda N and Suzuki M (2003) Malonylated flavonol glycosides from the petals of Clitoria ternatea. *Phytochemistry* 62:229-237.

Keyhanian S and Stahl-Biskup E (2007) Phenolic constituents in dried flowers of aloe vera (Aloe barbadensis) and their in vitro antioxidative capacity. *Planta medica* 73:599-602.

Kicel A and Olszewska MA (2015) Evaluation of antioxidant activity, and quantitative estimation of flavonoids, saponins and phenols in crude extract and dry fractions of Medicago lupulina aerial parts. *Natural product communications* 10:483-486.

Kicel A and Wolbiś M (2012) Study on the phenolic constituents of the flowers and leaves of Trifolium repens L. *Natural product research* 26:2050-2054.

Kim B-W, Lee E-R, Min H-M, Jeong H-S, Ahn J-Y, Kim J-H, Choi H-Y, Choi H, Kim EY and Park SP (2008) Sustained ERK activation is involved in the kaempferol-induced apoptosis of breast cancer cells and is more evident under 3-D culture condition. *Cancer biology & therapy* 7:1080-1089.

Kim JE, Jun HA, Cheigh HS and Choi JS (2002) Articles: A New Kaempferol 7-O-Triglucoside from the Leaves of Brassica juncea L. *김치의 과학과 기술* 8:73-76.

Kim JM, Lee EK, Kim DH, Yu BP and Chung HY (2010) Kaempferol modulates pro-inflammatory NF-κB activation by suppressing advanced glycation endproducts-induced NADPH oxidase. *Age* 32:197-208.

Kim Y, Park EJ, Kim J, Kim Y-B, Kim SR and Kim YC (2001) Neuroprotective constituents from Hedyotis diffusa. *Journal of natural products* 64:75-78.

Krasteva I, Bratkov V, Bucar F, Kunert O, Kollroser M, Kondeva-Burdina M and Ionkova I (2015) Flavoalkaloids and Flavonoids from Astragalus monspessulanus. *Journal of natural products* 78:2565-2571.

Krauze-Baranowska M, Sowiński P, Kawiak A and Sparzak B (2013) Flavonoids from Pseudotsuga menziesii. *Zeitschrift für Naturforschung C* 68:87-96.

Krenn L, Wollenweber E, Steyrleuthner K, Görick C and Melzig MF (2009) Contribution of methylated exudate flavonoids to the anti-inflammatory activity of Grindelia robusta. *Fitoterapia* 80:267-269.

Kumar M, Ahmad A, Rawat P, Khan MF, Rasheed N, Gupta P, Sathiamoorthy B, Bhatia G, Palit G and Maurya R (2010) Antioxidant flavonoid glycosides from Evolvulus alsinoides. *Fitoterapia* 81:234-242.

Kumar R, Ilyas M, Parveen M and Shafiullah (2006) A new chromone from Cassia nodosa. *Journal of Asian natural products research* 8:595-598.

Küpeli E, Orhan DD and Yesilada E (2006) Effect of Cistus laurifolius L. leaf extracts and flavonoids on acetaminophen-induced hepatotoxicity in mice. *Journal of ethnopharmacology* 103:455-460.

Küpeli E, Tatli II, Akdemir ZS and Yesilada E (2007) Estimation of antinociceptive and anti-inflammatory activity on Geranium pratense subsp. finitimum and its phenolic compounds. *Journal of ethnopharmacology* 114:234-240.

Kuti JO and Konuru HB (2004) Antioxidant capacity and phenolic content in leaf extracts of tree spinach (Cnidoscolus spp.). *Journal of Agricultural and Food Chemistry* 52:117-121.

Kwak JH, Kang MW, Roh JH, Choi SU and Zee OP (2009) Cytotoxic phenolic compounds from Chionanthus retusus. *Archives of pharmacal research* 32:1681-1687.

Kwon SH, Nam JI, Kim SH, Kim JH, Yoon JH and Kim KS (2009) Kaempferol and quercetin, essential ingredients in Ginkgo biloba extract, inhibit interleukin-1β-induced MUC5AC gene expression in human airway epithelial cells. *Phytotherapy Research* 23:1708-1712.

Landa A, Casado R and Calvo M (2009) Identification and quantification of flavonoids from Chuquiraga spinosa (Asteraceae). *Natural product communications* 4:1353-1355.

Lavola A, Karjalainen R and Julkunen-Tiitto R (2012) Bioactive polyphenols in leaves, stems, and berries of Saskatoon (Amelanchier alnifolia Nutt.) cultivars. *Journal of agricultural and food chemistry* 60:1020-1027.

Lee B, Kwon M, Choi JS, Jeong HO, Chung HY and Kim H-R (2015) Kaempferol Isolated from Nelumbo nucifera Inhibits Lipid Accumulation and Increases Fatty Acid Oxidation Signaling in Adipocytes. *Journal of medicinal food* 18:1363-1370.

Lee EH, Kim HJ, Song YS, Jin C, Lee K-T, Cho J and Lee YS (2003) Constituents of the stems and fruits ofOpuntia ficus-indica var. saboten. *Archives of pharmacal research* 26:1018-1023.

Lee JH, Ku CH, Baek N-l, Kim S-H, Park HW and Kim DK (2004) Phytochemical constituents from Diodia teres. *Archives of pharmacal research* 27:40-43.

Lee K-T, Choi J-H, Kim D-H, Son K-H, Kim W B, Kwon S-H and Park H-J (2001) Constitutents and the antitumor principle ofAllium victorialis var. platyphyllum. *Archives of pharmacal research* 24:44-50.

Lee MK, Jeon HY, Lee KY, Kim SH, Ma CJ, Sung SH, Lee H-S, Park MJ and Kim YC (2007) Inhibitory constituents of Euscaphis japonica on lipopolysaccharide-induced nitric oxide production in BV2 microglia. *Planta medica* 73:782-786.

Lee S-S, Chen W-C and Chen C H (2000) New jujubogenin glycosides from Colubrina asiatica. *Journal of natural products* 63:1580-1583.

Lee SY, Kim KH, Lee IK, Lee KH, Choi SU and Lee KR (2012) A new flavonol glycoside from Hylomecon vernalis. *Archives of pharmacal research* 35:415-421.

Lee VS-Y, Chen C-R, Liao Y-W, Tzen JT-C and Chang C-I (2008) Structural determination and DPPH radical-scavenging activity of two acylated flavonoid tetraglycosides in oolong tea (Camellia sinensis). *Chemical and Pharmaceutical Bulletin* 56:851-853.

Lei J, Xiao Y, Wang W, Xi Z, Liu M, Ran J and Huang J (2012) [Study on flavonoid chemical constituents contained in Memorialis hirta]. *Zhongguo*

Zhong yao za zhi= Zhongguo zhongyao zazhi= China journal of Chinese materia medica 37:478-482.

Lei J, Xiao Y, Wang W, Xi Z, Yu M and Huang J (2011) [Flavonoid glycosides from Androsace umbellata]. *Zhongguo Zhong yao za zhi= Zhongguo zhongyao zazhi= China journal of Chinese materia medica* 36:2353-2357.

Leite JPV, Rastrelli L, Romussi G, Oliveira AB, Vilegas JH, Vilegas W and Pizza C (2001) Isolation and HPLC quantitative analysis of flavonoid glycosides from Brazilian beverages (Maytenus ilicifolia and M. aquifolium). *Journal of agricultural and food chemistry* 49:3796-3801.

Leung HW-C, Lin C-J, Hour M-J, Yang W-H, Wang M-Y and Lee H-Z (2007) Kaempferol induces apoptosis in human lung non-small carcinoma cells accompanied by an induction of antioxidant enzymes. *Food and Chemical Toxicology* 45:2005-2013.

Li C-Y, Lee E-J and Wu T-S (2004) Antityrosinase Principles and Constituents of the Petals of Crocus s ativus. *Journal of natural products* 67:437-440.

Li D, Ikeda T, Matsuoka N, Nohara T, Zhang H, Sakamoto T and Nonaka G-I (2006a) Cucurbitane glycosides from unripe fruits of Lo Han Kuo (Siraitia grosvenori). *Chemical and pharmaceutical bulletin* 54:1425-1428.

Li H-f, Guan X-y, Yang W-z, Ye M, Sun C, Lu S and Guo D-a (2012) Antioxidant flavonoids from Epimedium wushanense. *Fitoterapia* 83:44-48.

Li H, Ji H-S, Kang J-H, Shin D-H, Park H-Y, Choi M-S, Lee C-H, Lee I-K, Yun B-S and Jeong T-S (2015) Soy Leaf Extract Containing Kaempferol Glycosides and Pheophorbides Improves Glucose Homeostasis by Enhancing Pancreatic β-Cell Function and Suppressing Hepatic Lipid Accumulation in db/db Mice. *Journal of agricultural and food chemistry* 63:7198-7210.

Li J-R, Liu J, He D-H, Xu H-X, Ding L-S, Bao W-K, Zhou Z-Q and Zhou Y (2013) Three new phenolic compounds from the leaves of Rosa sericea. *Fitoterapia* 84:332-337.

Li J, Huang H, Zhou W, Feng M and Zhou P (2008) Anti-hepatitis B virus activities of Geranium carolinianum L. extracts and identification of the active components. *Biological and Pharmaceutical Bulletin* 31:743-747.

Li M-M, Wang K, Pan Z-H, Chen X-Q, Peng L-Y, Li Y, Cheng X and Zhao Q-S (2009a) Two new sesquiterpene glucosides from Dennstaedtia scabra (Wall.) Moore. *Chemical and Pharmaceutical Bulletin* 57:1123-1125.

Li N, Li X, Meng D-L, Guo Y-Q and Wang J-H (2006b) Flavonoids from Camptosorus sibiricus Rupr. *Journal of Asian natural products research* 8:167-171.

Li W, Du B, Wang T, Wang S and Zhang J (2009b) Kaempferol induces apoptosis in human HCT116 colon cancer cells via the Ataxia-Telangiectasia Mutated-p53 pathway with the involvement of p53 Upregulated Modulator of Apoptosis. *Chemico-Biological Interactions* 177:121-127.

Li Y-F, Hu L-H, Lou F-C, Li J and Shen Q (2005) PTP1B inhibitors from Ardisia japonica. *Journal of Asian natural products research* 7:13-18.

Li Y, Fang H and Xu W (2007) Recent advance in the research of flavonoids as anticancer agents. *Mini reviews in medicinal chemistry* 7:663-678.

Liang D, Liu Y, Hao Z, Luo H, Wang Y, Zhang C, Zhang Q, Chen R and Yu D (2015) [Flavonol glycosides from Lysimachia clethroides]. *Zhongguo Zhong yao za zhi= Zhongguo zhongyao zazhi= China journal of Chinese materia medica* 40:103-107.

Liao Y, Fu H, Zhou G, Zhong R, Yuan H, Guo Q and Yan Q (2014) [Chemical Constituents from Leaves of Liquidambar formosana]. *Zhong yao cai= Zhongyaocai= Journal of Chinese medicinal materials* 37:2219-2221.

Lim Y-H, Kim I-H and Seo J-J (2007) In vitro activity of kaempferol isolated from the Impatiens balsamina alone and in combination with erythromycin or clindamycin against Propionibacterium acnes. *Journal of microbiology (Seoul, Korea)* 45:473-477.

Lin CW, Shen SC, Chien CC, Yang LY, Shia LT and Chen YC (2010) 12-O-tetradecanoylphorbol-13-acetate-induced invasion/migration of glioblastoma cells through activating PKCα/ERK/NF-κB-dependent MMP-9 expression. *Journal of cellular physiology* 225:472-481.

Lin Y-L, Shen C-C, Huang Y-J and Chang Y-Y (2005) Homoflavonoids from Ophioglossum p etiolatum. *Journal of natural products* 68:381-384.

Ling SK, Md. Pisar M, Man and Salbiah (2007) Platelet-activating factor (PAF) receptor binding antagonist activity of the methanol extracts and isolated flavonoids from Chromolaena odorata (L.) King and Robinson. *Biological and Pharmaceutical Bulletin* 30:1150-1152.

Liu A, Zou Z-M, Xu L-Z and Yang S-L (2005) A new cerebroside from Uvaria tonkinensis var. subglabra: Note. *Journal of Asian natural products research* 7:861-865.

Liu C, Sun H, Wang W, Sun J and Smith M (2015) [Chemical Constituents of Euphorbia lunulata]. *Zhong yao cai= Zhongyaocai= Journal of Chinese medicinal materials* 38:514-517.

Liu H, Gao Y, Wang K and Hu Z (2006) Determination of active components in Cynanchum chinense R. Br. by capillary electrophoresis. *Biomedical Chromatography* 20:451-454.

Liu H, Orjala J, Sticher O and Rali T (1999) Acylated flavonol glycosides from leaves of Stenochlaena palustris. *Journal of natural products* 62:70-75.

Liu J, Zhang X, Shi Y, Zhang Q, Ma Y and Chen J (2011a) [Studies on chemical constituents of Illicium simonsii]. *Zhongguo Zhong yao za zhi= Zhongguo zhongyao zazhi= China journal of Chinese materia medica* 36:1311-1315.

Liu Q, Ahn JH, Kim SB, Hwang BY and Lee MK (2012) New phenolic compounds with anti-adipogenic activity from the aerial parts of Pulsatilla koreana. *Planta medica* 78:1783-1786.

Liu X, Ye W, Yu B, Zhao S, Wu H and Che C (2004) Two new flavonol glycosides from Gymnema sylvestre and Euphorbia ebracteolata. *Carbohydrate research* 339:891-895.

Liu Y-M, Liu Q-H and Chen B-Q (2011b) A new flavonol glycoside from the seeds of Nigella glandulifera. *Natural product research* 25:1334-1338.

Loizzo MR, Said A, Tundis R, Hawas UW, Rashed K, Menichini F, Frega NG and Menichini F (2009) Antioxidant and antiproliferative activity of Diospyros lotus L. extract and isolated compounds. *Plant foods for human nutrition* 64:264-270.

Lopez-Lazaro M (2009) Distribution and biological activities of the flavonoid luteolin. *Mini reviews in medicinal chemistry* 9:31-59.

Luo H, Daddysman MK, Rankin GO, Jiang B-H and Chen YC (2010) Kaempferol enhances cisplatin's effect on ovarian cancer cells through promoting apoptosis caused by down regulation of cMyc. *Cancer cell international* 10:1.

Luo Y-G, Li B-G and Zhang G-L (2002) Four new glycosides from Pleurospermum franchetianum. *Journal of Asian natural products research* 4:155-163.

Luo Y, Deng Y, Chen B, Ding L and Wu FE (2006) A new amide from Thyrocarpus glochidiatus. *Natural product research* 20:1063-1066.

Lv X, Mai X, Guo H and Lai X (2012) [Chemical constituents of the roots of Vaccinium bracteatum]. *Zhong yao cai= Zhongyaocai= Journal of Chinese medicinal materials* 35:917-919.

Lv Z, Yin Y, Lin L and Peng Y (2014) [Chemical constituents from Canarium pimela fruits]. *Zhong yao cai= Zhongyaocai= Journal of Chinese medicinal materials* 37:1801-1803.

Ly TN, Shimoyamada M and Yamauchi R (2006) Isolation and characterization of rosmarinic acid oligomers in Celastrus hindsii Benth leaves and their antioxidative activity. *Journal of agricultural and food chemistry* 54:3786-3793.

Maciel MAM, Pinto AC, Arruda AC, Pamplona SG, Vanderlinde FA, Lapa AJ, Echevarria A, Grynberg NF, Côlus IM and Farias RA (2000) Ethnopharmacology, phytochemistry and pharmacology: a successful combination in the study of Croton cajucara. *Journal of Ethnopharmacology* 70:41-55.

Mahmoud AA, Al-Shihry SS and Hegazy M-EF (2009) Biological activity of a phloroglucinol glucoside derivative from Conyza aegyptiaca. *Zeitschrift für Naturforschung C* 64:513-517.

Mahmoud I, Moharram F, Marzouk M, Linscheid M and Saleh M (2002) Polyphenolic constituents of Callistemon lanceolatus leaves. *Die Pharmazie* 57:494-496.

Majinda R, Motswaledi M, Waigh RD and Waterman PG (1997) Phenolic and antibacterial constituents of Vahlia capensis. *Planta medica* 63:268-270.

Malafronte N, Pesca MS, Bisio A, Morales EL and De Tommasi N (2009) New flavonoid glycosides from Vernonia ferruginea. *Natural product communications* 4:1639-1642.

Mallavadhani UV, Narasimhan K, Sudhakar AVS, Mahapatra A, Li W and Breemen RBv (2006) Three new pentacyclic triterpenes and some flavonoids from the fruits of an Indian Ayurvedic plant Dendrophthoe falcata and their estrogen receptor binding activity. *Chemical and pharmaceutical bulletin* 54:740-744.

Manguro LOA, Ugi I, Hermann R and Lemmen P (2003a) Flavonol and drimane-type sesquiterpene glycosides of Warburgia stuhlmannii leaves. *Phytochemistry* 63:497-502.

Manguro LOA, Ugi I, Lemmen P and Hermann R (2003b) Flavonol glycosides of Warburgia ugandensis leaves. *Phytochemistry* 64:891-896.

Marín C, Boutaleb-Charki S, Diaz JG, Huertas O, Rosales MJ, Pérez-Cordon G, Guitierrez-Sánchez R and Sánchez-Moreno M (2009) Antileishmaniasis activity of flavonoids from Consolida oliveriana. *Journal of natural products* 72:1069-1074.

Marino A, Zengin G, Nostro A, Ginestra G, Dugo P, Cacciola F, Miceli N, Taviano MF, Filocamo A and Bisignano G (2016) Antimicrobial activities, toxicity and phenolic composition of Asphodeline anatolica E. Tuzlaci leaf extracts from Turkey. *Natural product research*:1-4.

Marniemi J, Alanen E, Impivaara O, Seppänen R, Hakala P, Rajala T and Rönnemaa T (2005) Dietary and serum vitamins and minerals as predictors of myocardial infarction and stroke in elderly subjects. *Nutrition, metabolism and cardiovascular diseases* 15:188-197.

Maron DJ (2004) Flavonoids for reduction of atherosclerotic risk. *Current atherosclerosis reports* 6:73-78.

Marzouk MS, El-Toumy SA, Moharram FA, Shalaby NM and Ahmed AA (2002) Pharmacologically active ellagitannins from Terminalia myriocarpa. *Planta medica* 68:523-527.

Meng Z, Zhou Y, Lu J, Sugahara K, Xu S and Kodama H (2001) Effect of five flavonoid compounds isolated from Quercus dentata Thunb on superoxide generation in human neutrophils and phosphorylation of neutrophil proteins. *Clinica chimica acta* 306:97-102.

Meyre-Silva C, Yunes RA, Monache F, Santos ARS, Schmeling LO, de Maria Gadotti V, Liz F and Cechinel-Filho V (2001) Phytochemical and pharmacological analysis of Bauhinia microstachya (Raddi) Macbr.(Leguminosae). *ZEITSCHRIFT FUR NATURFORSCHUNG C* 56:939-942.

Min B-S, Tomiyama M, Ma C-m, Nakamura N and Hattori M (2001) Kaempferol acetylrhamnosides from the rhizome of Dryopteris crassirhizoma and their inhibitory effects on three different activities of human immunodeficiency virus-1 reverse transcriptase. *Chemical and pharmaceutical bulletin* 49:546-550.

Mira A, Yamashita S, Katakura Y and Shimizu K (2015) In Vitro Neuroprotective Activities of Compounds from Angelica shikokiana Makino. *Molecules* 20:4813-4832.

Mirza B and Ahmad MS (2015) Evaluation of antioxidant potential and HPLC based identification of phenolics in Polygonum amplexicaule extract and its fractions. *Pak J Pharm Sci* 28:431-435.

Mizuno M, Kyotani Y, Iinuma M, Tanaka T and Iwatsuki K (1990) Kaempferol 3-rhamnoside-7-[6-feruloylglucosyl (1→ 3) rhamnoside] from Asplenium prolongatum. *Phytochemistry* 29:2742-2743.

Mokhtar M, Soukup J, Donato P, Cacciola F, Dugo P, Riazi A, Jandera P and Mondello L (2015) Determination of the polyphenolic content of a Capsicum annuum L. extract by liquid chromatography coupled to photodiode array and mass spectrometry detection and evaluation of its biological activity. *Journal of separation science* 38:171-178.

Mokhtari-Zaer A, Khazdair MR and Boskabady MH (2015) Smooth muscle relaxant activity of Crocus sativus (saffron) and its constituents: possible mechanisms. *Avicenna journal of phytomedicine* 5:365.

Moon S-S, Rahman AA, Manir M and Ahamed VJ (2010) Kaempferol glycosides and cardenolide glycosides, cytotoxic constituents from the seeds of Draba nemorosa (Brassicaceae). *Archives of pharmacal research* 33:1169-1173.

Moreno A, Martín-Cordero C, Iglesias-Guerra F and Toro MV (2002) Flavonoids from Dorycnium rectum. *Biochemical systematics and ecology* 30:73-74.

Moriyama H, Iizuka T, Nagai M, Miyataka H and Satoh T (2003) Antiinflammatory activity of heat-treated Cassia alata leaf extract and its flavonoid glycoside. *Yakugaku Zasshi* 123:607-611.

Mostafa NM, Ashour ML, Eldahshan OA and Singab ANB (2015) Cytotoxic activity and molecular docking of a novel biflavonoid isolated from Jacaranda acutifolia (Bignoniaceae). *Natural product research*:1-8.

Murakami T, Kohno K, Kishi A, MATSUDA H and YOSHIKAWA M (2000) Medicinal foodstuffs. XIX. Absolute stereostructures of canavalioside, a new Ent-kaurane-type diterpene glycoside, and gladiatosides A1, A2, A3, B1, B2, B3, C1, and C2, new acylated flavonol glycosides, from sword bean, the seeds of Canavalia gladiata. *Chemical and pharmaceutical bulletin* 48:1673-1680.

Mussadiq S, Riaz N, Saleem M, Ashraf M, Ismail T and Jabbar* A (2013) New acylated flavonoid glycosides from flowers of Aerva javanica. *Journal of Asian natural products research* 15:708-716.

Muthukrishnan SD, Kaliyaperumal A and Subramaniyan A (2015) Identification and determination of flavonoids, carotenoids and chlorophyll concentration in Cynodon dactylon (L.) by HPLC analysis. *Natural product research* 29:785-790.

Muzitano MF, Tinoco LW, Guette C, Kaiser CR, Rossi-Bergmann B and Costa SS (2006) The antileishmanial activity assessment of unusual flavonoids from Kalanchoe pinnata. *Phytochemistry* 67:2071-2077.

Mylonis I, Lakka A, Tsakalof A and Simos G (2010) The dietary flavonoid kaempferol effectively inhibits HIF-1 activity and hepatoma cancer cell viability under hypoxic conditions. *Biochem Bioph Res Co* 398:74-78.

Nawwar M, Ayoub N, El-Raey M, Zaghloul S, Hashem A, Mostafa E, Eldahshan O, Lindequist U and Linscheid MW (2015) Acylated flavonol diglucosides from Ammania auriculata. *Zeitschrift für Naturforschung C* 70:39-43.

Nazaruk J and Jakoniuk P (2005) Flavonoid composition and antimicrobial activity of Cirsium rivulare (Jacq.) All. flowers. *Journal of ethnopharmacology* 102:208-212.

Nazemiyeh H, Kazemi EM, Zare K, Jodari M, Nahar L and Sarker SD (2010) Free radical scavengers from the aerial parts of Euphorbia petiolata. *Journal of natural medicines* 64:187-190.

Neuhouser ML (2004) Review: Dietary flavonoids and cancer risk: Evidence from human population studies. *Nutrition and cancer* 50:1-7.

Nguemeving JR, Azebaze AGB, Kuete V, Carly NNE, Beng VP, Meyer M, Blond A, Bodo B and Nkengfack AE (2006) Laurentixanthones A and B, antimicrobial xanthones from Vismia laurentii. *Phytochemistry* 67:1341-1346.

Nielsen JK, Olsen CE and Petersen MK (1993) Acylated flavonol glycosides from cabbage leaves. *Phytochemistry* 34:539-544.

Niklas KJ and Giannasi DE (1977) Flavonoids and other chemical constituents of fossil Miocene Zelkova (Ulmaceae). *Science* 196:877-878.

Ning D, Yan X, Huang S, Cheng L, Li J and Pan Z (2015) [Studies on chemical constituents of Zhuang medicine Excoecaria venenata and their cytotoxic activity]. *Zhongguo Zhong yao za zhi= Zhongguo zhongyao zazhi= China journal of Chinese materia medica* 40:686-690.

Nørbæk R and Kondo T (1999) Flavonol glycosides from flowers of Crocus speciosus and C. antalyensis. *Phytochemistry* 51:1113-1119.

Nöthlings U, Murphy SP, Wilkens LR, Henderson BE and Kolonel LN (2007) Flavonols and pancreatic cancer risk the multiethnic cohort study. *American journal of epidemiology* 166:924-931.

Nowak R and Gawlik-Dziki U (2007) Polyphenols of Rosa L. leaves extracts and their radical scavenging activity. *Zeitschrift für Naturforschung C* 62:32-38.

Ntandou GN, Banzouzi J, Mbatchi B, Elion-Itou R, Etou-Ossibi A, Ramos S, Benoit-Vical F, Abena A and Ouamba J (2010) Analgesic and anti-inflammatory effects of Cassia siamea Lam. stem bark extracts. *Journal of ethnopharmacology* 127:108-111.

Obmann A, Tsendayush D, Thalhammer T, Zehl M, Vo TPN, Purevsuren S, Natsagdorj D, Narantuya S, Kletter C and Glasl S (2010) Extracts from the Mongolian traditional medicinal plants Dianthus versicolorFisch. and Lilium pumilum Delile stimulate bile flow in an isolated perfused rat liver model. *Journal of ethnopharmacology* 131:555-561.

Odabas MS, Camas N, Cirak C, Radusiene J, Janulis V and Ivanauskas L (2010) The quantitative effects of temperature and light intensity on phenolics accumulation in St. John's wort (Hypericum perforatum). *Natural product communications* 5:535-540.

Oh H, Kang D-G, Kwon J-W, Kwon T-O, Lee S-Y, Lee D-B and Lee H-S (2004a) Isolation of angiotensin converting enzyme (ACE) inhibitory flavonoids from Sedum sarmentosum. *Biological and Pharmaceutical Bulletin* 27:2035-2037.

Oh H, Kim D-H, Cho J-H and Kim Y-C (2004b) Hepatoprotective and free radical scavenging activities of phenolic petrosins and flavonoids isolated from Equisetum arvense. *Journal of Ethnopharmacology* 95:421-424.

Okoye FB, Sawadogo WR, Sendker J, Aly AH, Quandt B, Wray V, Hensel A, Esimone CO, Debbab A and Diederich M (2015) Flavonoid glycosides from Olax mannii: Structure elucidation and effect on the nuclear factor kappa B pathway. *Journal of ethnopharmacology* 176:27-34.

Öksüz S, Gürek F, Lin L-z, Gil RR, Pezzuto JM and Cordell GA (1996) Aleppicatines A and B from Euphorbia aleppica. *Phytochemistry* 42:473-478.

Olsen H, Aaby K and Borge GIA (2009) Characterization and quantification of flavonoids and hydroxycinnamic acids in curly kale (Brassica oleracea L. convar. acephala var. sabellica) by HPLC-DAD-ESI-MS n. *Journal of agricultural and food chemistry* 57:2816-2825.

Orhan I, Küpeli E, Terzioğlu S and Yesilada E (2007) Bioassay-guided isolation of kaempferol-3-O-β-d-galactoside with anti-inflammatory and antinociceptive activity from the aerial part of Calluna vulgaris L. *Journal of ethnopharmacology* 114:32-37.

Oszmianski J, Kolniak-Ostek J and Wojdyło A (2012) Characterization and content of flavonol derivatives of Allium ursinum L. plant. *Journal of agricultural and food chemistry* 61:176-184.

Özipek M, Çaliş İ, Ertan M and Rüedi P (1994) Rhamnetin 3-p-coumaroylrhamninoside from Rhamnus petiolaris. *Phytochemistry* 37:249-253.

Papiez M, Gancarczyk M and Bilińska B (2001) The compounds from the hollyhock extract (Althaea rosea Cav. var. nigra) affect the aromatization in rat testicular cells in vivo and in vitro. *Folia histochemica et cytobiologica/Polish Academy of Sciences, Polish Histochemical and Cytochemical Society* 40:353-359.

Parejo I, Viladomat F, Bastida J, Schmeda-Hirschmann G, Burillo J and Codina C (2004) Bioguided isolation and identification of the nonvolatile antioxidant compounds from fennel (Foeniculum vulgare Mill.) waste. *Journal of agricultural and food chemistry* 52:1890-1897.

Park CH, Kim KH, Lee IK, Lee SY, Choi SU, Lee JH and Lee KR (2011) Phenolic constituents of Acorus gramineus. *Archives of pharmacal research* 34:1289-1296.

Pattamadilok D and Suttisri R (2008) Seco-terpenoids and other constituents from Elateriospermum tapos. *Journal of natural products* 71:292-294.

Pistelli L, Noccioli C, Giachi I, Dimitrova B, Gevrenova R, Morelli I and Potenza D (2005) Lupane-triterpenes from Bupleurum flavum. *Natural product research* 19:783-788.

Prawat H, Mahidol C, Ruchirawat S, Prawat U, Tuntiwachwuttikul P, Tooptakong U, Taylor WC, Pakawatchai C, Skelton BW and White AH (1995) Cyanogenic and non-cyanogenic glycosides from Manihot esculenta. *Phytochemistry* 40:1167-1173.

Qin L, Ming L, Mabry TJ and Dixon RA (1994) Flavonol glycosides from Cephalocereus senilis. *Phytochemistry* 36:229-231.

Qiu Y, Chen Y, Pei Y, Matsuda H and Yoshikawa M (2002) Constituents with Radical Scavenging Effect from Opuntia dillenii: Structures of New. ALPHA.-Pyrones and Flavonol Glycoside. *Chemical and pharmaceutical bulletin* 50:1507-1510.

Qu Q, Zhang L, Bao H, Zhang J, You X and Wang J (2011) [Chemical constituents of flavonoids from flowers of Koelreuteria paniculata]. *Zhong yao cai= Zhongyaocai= Journal of Chinese medicinal materials* 34:1716-1719.

Quintyne-Walcott S, Maxwell AR and Reynolds WF (2007) Crotogossamide, a cyclic nonapeptide from the latex of Croton gossypifolius. *Journal of natural products* 70:1374-1376.

Qureshi MN, Stecher G and Bonn GK (2014) Determination of total polyphenolic compounds and flavonoids in Juglans regia leaves. *Pakistan journal of pharmaceutical sciences* 27:865-869.

Ramadan M, Ahmad A, Nafady A and Mansour A (2009) Chemical composition of the stem bark and leaves of Ficus pandurata Hance. *Natural product research* 23:1218-1230.

Rao YK, Fang S-H and Tzeng Y-M (2008) Antiinflammatory activities of flavonoids and a triterpene caffeate isolated from Bauhinia variegata. *Phytotherapy research* 22:957-962.

Rashed KN and Butnariu M (2014) Isolation and antimicrobial and antioxidant evaluation of bio-active compounds from Eriobotrya japonica stems. *Advanced pharmaceutical bulletin* 4:75.

Regasini LO, Vellosa JCR, Silva DHS, Furlan M, de Oliveira OMM, Khalil NM, Brunetti IL, Young MCM, Barreiro EJ and Bolzani VS (2008)

Flavonols from Pterogyne nitens and their evaluation as myeloperoxidase inhibitors. *Phytochemistry* 69:1739-1744.

Ribeiro L, Ribani R, Francisco T, Soares A, Pontarolo R and Haminiuk C (2015) Profile of bioactive compounds from grape pomace (Vitis vinifera and Vitis labrusca) by spectrophotometric, chromatographic and spectral analyses. *Journal of Chromatography B* 1007:72-80.

Rocha L, Marston A, Potterat O, Kaplan MAC, Stoeckli-Evans H and Hostettmann K (1995) Antibacterial phloroglucinols and flavonoids from Hypericum brasiliense. *Phytochemistry* 40:1447-1452.

Rocha RS, Kassuya CAL, Formagio ASN, Mauro MdO, Andrade-Silva M, Monreal ACD, Cunha-Laura AL, Vieira MdC and Oliveira RJ (2015) Analysis of the anti-inflammatory and chemopreventive potential and description of the antimutagenic mode of action of the Annona crassiflora methanolic extract. *Pharmaceutical biology*:1-13.

Rochfort SJ, Imsic M, Jones R, Trenerry VC and Tomkins B (2006) Characterization of flavonol conjugates in immature leaves of pak choi [Brassica rapa L. Ssp. chinensis L.(Hanelt.)] by HPLC-DAD and LC-MS/MS. *Journal of agricultural and food chemistry* 54:4855-4860.

Rodríguez Galdón B, Rodríguez Rodríguez E and Díaz Romero C (2008) Flavonoids in onion cultivars (Allium cepa L.). *Journal of food science* 73:C599-C605.

Rodríguez P, González-Mujica F, Bermúdez J and Hasegawa M (2010) Inhibition of glucose intestinal absorption by kaempferol 3-O-α-rhamnoside purified from Bauhinia megalandra leaves. *Fitoterapia* 81:1220-1223.

Ross SA, ElSohly MA, Sultana GN, Mehmedic Z, Hossain CF and Chandra S (2005) Flavonoid Glycosides and Cannabinoids from the Pollen of Cannabis sativa L. *Phytochemical Analysis* 16:45-48.

Ruan S, Lu Y and Chen D (2013) [Anti-complementary constituents of Pogostemon cablin]. *Zhongguo Zhong yao za zhi= Zhongguo zhongyao zazhi= China journal of Chinese materia medica* 38:2129-2135.

Ryan KG, Swinny EE, Winefield C and Markham KR (2001) Flavonoids and UV photoprotection in Arabidopsis mutants. *Zeitschrift für Naturforschung C* 56:745-754.

Rzadkowska-Bodalska H, Olechnowicz-Stepien W and Rzepka A (1974) The components of ethereal and methanolic extracts from the fruits of Celastrus tatarinovii (Rupr.). *Polish journal of pharmacology and pharmacy* 27:335-338.

Saito N, Tatsuzawa F, Toki K, Shinoda K, Shigihara A and Honda T (2011) The blue anthocyanin pigments from the blue flowers of Heliophila coronopifolia L.(Brassicaceae). *Phytochemistry* 72:2219-2229.

Sajeli Begum A, Sahai M, Fujimoto Y, Asai K, Schneider K, Nicholson G and Suessmuth R (2006) A new kaempferol diglycoside from Datura suaveolens Humb. & Bonpl. ex. Willd. *Natural product research* 20:1231-1236.

Sala A, Recio MC, Schinella GR, Máñez S, Giner RM, Cerdá-Nicolás M and Ríos J-L (2003) Assessment of the anti-inflammatory activity and free radical scavenger activity of tiliroside. *European Journal of Pharmacology* 461:53-61.

Salvador MJ, Ferreira EO, Mertens-Talcott SU, De Castro WV, Butterweck V, Derendorf H and Dias DA (2006) Isolation and HPLC quantitative analysis of antioxidant flavonoids from Alternanthera tenella Colla. *Zeitschrift für Naturforschung C* 61:19-25.

Samy MN, Hamed ANE-S, Sugimoto S, Otsuka H, Kamel MS and Matsunami K (2015) Officinalioside, a new lignan glucoside from Borago officinalis L. *Natural product research*:1-6.

Satake T, Kamiya K, An Y, Oishi T and Yamamoto J (2007) The anti-thrombotic active constituents from Centella asiatica. *Biological and Pharmaceutical Bulletin* 30:935-940.

Schmitzer V, Veberic R, Slatnar A and Stampar F (2010) Elderberry (Sambucus nigra L.) wine: a product rich in health promoting compounds. *Journal of agricultural and food chemistry* 58:10143-10146.

Scio E, Ribeiro A, Alves TM, Romanha AJ, de Souza Filho JD, Cordell GA and Zani CL (2003) Diterpenes from Alomia myriadenia (Asteraceae) with cytotoxic and trypanocidal activity. *Phytochemistry* 64:1125-1131.

Seetharaman T and Petrus A (2004) Note: Novel acylkaempferol glycoside from the endemic species, Vernonia travancorica Hook. f. *Journal of Asian natural products research* 6:295-299.

Semmar N, Fenet B, Gluchoff-Fiasson K, Hasan A and Jay M (2002) Four New Flavonol Glycosides from the Leaves of Astragalus c aprinus. *Journal of natural products* 65:576-579.

Semwal S, Sharma P, Kumar M and Sharma RK (2015) Chcmical constituents from rhizomes of Cautleya spicata (Sm.) Baker (Zingiberaceae). *Natural product research*:1-3.

Shao M, Wang Y, Jian Y, Sun X, Huang X, Zhang X and Ye W (2014) [Chemical constituents of leaves of Psidium guajava]. *Zhongguo Zhong yao za zhi= Zhongguo zhongyao zazhi= China journal of Chinese materia medica* 39:1024-1029.

Sharaf M (2008) Chemical constituents from the seeds of Trifolium alexandrinum. *Natural product research* 22:1620-1623.

Sharififar F, Yassa N and Mozaffarian V (2010) Bioactivity of major components from the seeds of Bunium persicum (Boiss.) Fedtch. *Pak J Pharm Sci* 23:300-304.

Sharma UK, Sharma K, Sharma N, Sharma A, Singh HP and Sinha AK (2007) Microwave-assisted efficient extraction of different parts of Hippophae rhamnoides for the comparative evaluation of antioxidant activity and quantification of its phenolic constituents by reverse-phase high-performance liquid chromatography (RP-HPLC)†. *Journal of agricultural and food chemistry* 56:374-379.

Sikorska M (2003) Flavonoids in the leaves of Asclepias incarnata L. *Acta poloniae pharmaceutica* 60:471-476.

Sikorska M, Matlawska I and Franski R (2001) Kaempferol and its glycosides in the seeds hair of Asclepias syriaca L. *Acta poloniae pharmaceutica* 58:211-216.

Singab AnB (1998) Acetylated flavonol triglycosides from Ammi majus L. *Phytochemistry* 49:2177-2180.

Singh S, Pandey M, Singh A, Singh U and Pandey V (2008) A new chalcone glycoside from Rhamnus nipalensis. *Natural product research* 22:1657-1659.

Song EK, Kim JH, Kim JS, Cho H, Nan JX, Sohn DH, Ko GI, Oh H and Kim YC (2003) Hepatoprotective phenolic constituents of Rhodiola sachalinensis on tacrine-induced cytotoxicity in Hep G2 cells. *Phytotherapy Research* 17:563-565.

Song J, Yang Y, Qi H and Li Q (2013) [Chemical constituents from flowers of Gardenia jasminoides]. *Zhong yao cai= Zhongyaocai= Journal of Chinese medicinal materials* 36:752-755.

Sosa T, Chaves N, Alias J, Escudero J, Henao F and Gutiérrez-Merino C (2004) Inhibition of mouth skeletal muscle relaxation by flavonoids of Cistus ladanifer L.: a plant defense mechanism against herbivores. *Journal of chemical ecology* 30:1087-1101.

Srivastava N, Srivastava A, Srivastava S, Rawat AKS and Khan AR (2015) Simultaneous Quantification of Syringic Acid and Kaempferol in Extracts of Bergenia Species Using Validated High-Performance Thin-Layer Chromatographic-Densitometric Method. *Journal of chromatographic science*:bmv154.

Stintzing FC, Kammerer D, Schieber A, Adama H, Nacoulma OG and Carle R (2004) Betacyanins and Phenolic Compounds from Amaranthus spinosus L. and Boerhavia erecta L. *Zeitschrift für Naturforschung C* 59:1-8.

Stojković D, Barros L, Petrović J, Glamoclija J, Santos-Buelga C, Ferreira IC and Soković M (2015) Ethnopharmacological uses of Sempervivum tectorum L. in southern Serbia: Scientific confirmation for the use against otitis linked bacteria. *Journal of ethnopharmacology* 176:297-304.

Su D, Tang W, Yu S, Liu Y, Qu J and Yu D (2012) [Chemical constituents of Neoalsomitra integrifoliola]. *Zhongguo Zhong yao za zhi= Zhongguo zhongyao zazhi= China journal of Chinese materia medica* 37:1593-1596.

Sumino M, Sekine T, Ruangrungsi N, Igarashi K and Ikegami F (2002) Ardisiphenols and other antioxidant principles from the fruits of Ardisia colorata. *Chemical and Pharmaceutical Bulletin* 50:1484-1487.

Sun L, Mu S, Wei L, Xu C and Jia M (2015) [Simultaneous determination of four flavonoids in Wikstroemia indica by HPLC]. *Zhongguo Zhong yao za zhi= Zhongguo zhongyao zazhi= China journal of Chinese materia medica* 40:700-703.

Suntornsuk L and Anurukvorakun O (2005) Precision improvement for the analysis of flavonoids in selected Thai plants by capillary zone electrophoresis. *Electrophoresis* 26:648-660.

Swinny EE (2001) A novel acetylated 3-deoxyanthocyanidin laminaribioside from the fern Blechnum novae-zelandiae. *Zeitschrift für Naturforschung C* 56:177-180.

Taskova R, Mitova M, Mikhova B and Duddeck H (2003) Bioactive phenolics from Carthamus lanatus L. *Zeitschrift für Naturforschung C* 58:704-707.

Terreaux C, Wang Q, Ioset J-R, Ndjoko K, Grimminger W and Hostettmann K (2002) Complete LC/MS analysis of a Tinnevelli senna pod extract and subsequent isolation and identification of two new benzophenone glucosides. *Planta medica* 68:349-354.

Tomaino A, Martorana M, Arcoraci T, Monteleone D, Giovinazzo C and Saija A (2010) Antioxidant activity and phenolic profile of pistachio (Pistacia vera L., variety Bronte) seeds and skins. *Biochimie* 92:1115-1122.

Tomczyk M, Gudej J and Sochacki M (2002) Flavonoids from Ficaria verna Huds. *Zeitschrift für Naturforschung C* 57:440-444.

Trinh PTN, Tri MD, Hien DC, An NH, Minh PN, An PN and Dung LT (2015) A new flavan from the Drynaria bonii H. Christ rhizomes. *Natural product research*:1-7.

Ueda Y, Oku H, Iinuma M and Ishiguro K (2003) Effects on blood pressure decrease in response to PAF of Impatiens textori MIQ. *Biological and Pharmaceutical Bulletin* 26:1505-1507.

Umehara K, Nemoto K, Ohkubo T, Miyase T, Degawa M and Noguchi H (2004) Isolation of a new 15-membered macrocyclic glycolipid lactone, Cuscutic Resinoside a from the seeds of Cuscuta chinensis: a stimulator of breast cancer cell proliferation. *Planta medica* 70:299-304.

Vachálková A, Eisenreichová E, Haladová M, Mucaji P, Jozova B and Novotný L (1999) Potential carcinogenic and inhibitory activity of compounds isolated from Lilium candidum L. *Neoplasma* 47:313-318.

Van Elswijk DA, Schobel UP, Lansky EP, Irth H and van der Greef J (2004) Rapid dereplication of estrogenic compounds in pomegranate (Punica granatum) using on-line biochemical detection coupled to mass spectrometry. *Phytochemistry* 65:233-241.

Vareed SK, Schutzki RE and Nair MG (2007) Lipid peroxidation, cyclooxygenase enzyme and tumor cell proliferation inhibitory compounds in Cornus kousa fruits. *Phytomedicine* 14:706-709.

Veit M and Pauli GF (1999) Major flavonoids from Arabidopsis thaliana leaves. *Journal of Natural Products* 62:1301-1303.

Vilegas W, Sanommiya M, Rastrelli L and Pizza C (1999) Isolation and structure elucidation of two new flavonoid glycosides from the infusion of Maytenus aquifolium leaves. Evaluation of the antiulcer activity of the infusion. *Journal of agricultural and food chemistry* 47:403-406.

Viola H, Wolfman C, De Stein ML, Wasowski C, Pena C, Medina J and Paladini A (1994) Isolation of pharmacologically active benzodiazepine receptor ligands from Tilia tomentosa (Tiliaceae). *Journal of ethnopharmacology* 44:47-53.

Vitalini S, Braca A and Fico G (2011) Study on secondary metabolite content of Helleborus niger L. leaves. *Fitoterapia* 82:152-154.

Wang G-J, Tsai T-H and Lin L-C (2007) Prenylflavonol, acylated flavonol glycosides and related compounds from Epimedium sagittatum. *Phytochemistry* 68:2455-2464.

Wang G, Lin B, Bai Y and Qin M (2012) [Chemical constituents from Gueldenstaedtia stenophylla]. *Zhongguo Zhong yao za zhi= Zhongguo zhongyao zazhi= China journal of Chinese materia medica* 37:1412-1416.

Wang L, Yu M, Chi Y, Ouyang W, Zang Z and Zhao Y (2014) [Chemical constituents of Euphorbia dracunculoides]. *Zhongguo Zhong yao za zhi= Zhongguo zhongyao zazhi= China journal of Chinese materia medica* 39:3969-3973.

Wang Q and Xu F (2013) [Chemical constituents of ethyl acetate fraction from Hypericum hengshanense]. *Zhong yao cai= Zhongyaocai= Journal of Chinese medicinal materials* 36:1611-1613.

Wang W-S, Lu P, Duan C-H and Feng J-C (2010) A new jacaranone derivative from Senecio scandens var. incisus. *Natural product research* 24:370-374.

Wang X-L, Wang N-L, Zhang Y, Gao H, Pang W-Y, Wong M-S, Zhang G, Qin L and Yao X-S (2008) Effects of eleven flavonoids from the osteoprotective fraction of Drynaria fortunei (KUNZE) J. SM. on osteoblastic proliferation using an osteoblast-like cell line. *Chemical and Pharmaceutical Bulletin* 56:46-51.

Wang YF, Cao JX, Efferth T, Lai GF and Luo SD (2006) Cytotoxic and new tetralone derivatives from Berchemia floribunda (Wall.) Brongn. *Chemistry & biodiversity* 3:646-653.

Wei B-L, Lu C-M, Tsao L-T, Wang J-P and Lin C-N (2001) In vitro anti-inflammatory effects of quercetin 3-O-methyl ether and other constituents from Rhamnus species. *Planta medica* 67:745-747.

Wei J, Chang X, Wang W and Kang W (2013) [Chemical constituents from Lysimachia circaeoides]. *Zhong yao cai= Zhongyaocai= Journal of Chinese medicinal materials* 36:1441-1443.

Williams CA, Harborne JB, Newman M, Greenham J and Eagles J (1997) Chrysin and other leaf exudate flavonoids in the genus Pelargonium. *Phytochemistry* 46:1349-1353.

Wirasathien L, Pengsuparp T, Moriyasu M, Kawanishi K and Suttisri R (2006) Cytotoxic C-benzylated chalcone and other constituents ofEllipeiopsis cherrevensis. *Archives of pharmacal research* 29:497-502.

Wong K-C, Hag Ali DM and Boey P-L (2012) Chemical constituents and antibacterial activity of Melastoma malabathricum L. *Natural product research* 26:609-618.

Wu B, Takahashi T, Kashiwagi T, Tebayashi S-i and Kim C-S (2007) New flavonoid glycosides from the leaves of Solidago altissima. *Chemical and pharmaceutical bulletin* 55:815-816.

Wu S, Sun Q, Chu C and Zhang J (2012a) [Chemical constituents of Eupatorium lindleyanum]. *Zhongguo Zhong yao za zhi= Zhongguo zhongyao zazhi= China journal of Chinese materia medica* 37:937-940.

Wu Y-Q, Li Y, Lin X-F, Liu Z-L and Li Y (2003) Phenolic compounds from Anaphalis aureo-punctata. *Die Pharmazie-An International Journal of Pharmaceutical Sciences* 58:833-835.

Wu Y, Luo Q, Sun C, Wang G, Chen Q, Guo Z, Zou X and Chen H (2012b) [Chemical constituents contained in Desmodium caudatum]. *Zhongguo Zhong yao za zhi= Zhongguo zhongyao zazhi= China journal of Chinese materia medica* 37:1788-1792.

Xiang W, Li R-T, Mao Y-L, Zhang H-J, Li S-H, Song Q-S and Sun H-D (2005) Four new prenylated isoflavonoids in Tadehagi triquetrum. *Journal of agricultural and food chemistry* 53:267-271.

Xiao K, Lai R, Lin R, Wu G, Zheng W and Wu D (2014) [Effects of different culture conditions on main chemical compositions of Anoectochilus roxburghii]. *Zhong yao cai= Zhongyaocai= Journal of Chinese medicinal materials* 37:553-556.

Xiao M, Cao N, Fan J, Shen Y and Xu Q (2011) [Studies on flavonoids from the leaves of Lindera aggregata]. *Zhong yao cai= Zhongyaocai= Journal of Chinese medicinal materials* 34:62-64.

Xie D, Li Y, Zhao L, Ding G, Yuan S, Xu J, Zhu H and Xiao W (2011) [Study on chemical constituents from leaves of Naudea officinalis]. *Zhongguo Zhong yao za zhi= Zhongguo zhongyao zazhi= China journal of Chinese materia medica* 36:1037-1039.

Xie J-H, Dong C-j, Nie S-P, Li F, Wang Z-J, Shen M-Y and Xie M-Y (2015) Extraction, chemical composition and antioxidant activity of flavonoids from Cyclocarya paliurus (Batal.) Iljinskaja leaves. *Food chemistry* 186:97-105.

Xin H-L, Wu Y-C, Su Y-H, Sheng J-Y and Ling C-Q (2011) Novel flavonoids from the leaves of Actinidia valvata Dunn: structural elucidation and antioxidant activity. *Planta medica* 77:70-73.

Xiong Q, Fan W, Tezuka Y, Adnyana IK, Stampoulis P, Hattori M, Namba T and Kadota S (2000) Hepatoprotective effect of Apocynum venetum and its active constituents. *Planta medica* 66:127-133.

Xu L-R, Zhou P, Zhi Y-E, Wu J and Zhang S (2009) Three new flavonol triglycosides from Derris trifoliata. *Journal of Asian natural products research* 11:79-84.

Xu L, Li Y, Li C, Lin L, Wang Z and Luo Y (2013) [Chemical constituents from whole plants of Aconitum tanguticum]. *Zhongguo Zhong yao za zhi= Zhongguo zhongyao zazhi= China journal of Chinese materia medica* 38:2818-2825.

Xu Y-m, Smith JA, Lannigan DA and Hecht SM (2006) Three acetylated flavonol glycosides from Forsteronia refracta that specifically inhibit p90 RSK. *Bioorganic & medicinal chemistry* 14:3974-3977.

Yahara S, Kohjyouma M and Kohoda H (2000) Flavonoid glycosides and saponins from Astragalus shikokianus. *Phytochemistry* 53:469-471.

Yan L-H, Liu X-Q, Zhu H, Xu Q-R, Wang W-M, Zhang S-M, Zhang Q-W, Zhang S-S and Wang Z-M (2015) Chemical constituents of Euonymus fortunei. *Journal of Asian natural products research*:1-7.

Yang D, Wei J, Yang Y, Yang Y and Li X (2013) [Chemical constituents of Euphorbia sikkimensis]. *Zhongguo Zhong yao za zhi= Zhongguo zhongyao zazhi= China journal of Chinese materia medica* 38:4094-4098.

Yang F, Yue Z, Wang X, Zhang X, Chai J, Cui J, Song X and Mei Q (2014) [Chemical constituents of leaf of Eucommia ulmoides]. *Zhongguo Zhong yao za zhi= Zhongguo zhongyao zazhi= China journal of Chinese materia medica* 39:1445-1449.

Yang H, Protiva P, Cui B, Ma C, Baggett S, Hequet V, Mori S, Weinstein IB and Kennelly EJ (2003) New Bioactive Polyphenols from Theobroma g randiflorum ("Cupuaçu"). *Journal of natural products* 66:1501-1504.

Yang R-Y, Lin S and Kuo G (2008) Content and distribution of flavonoids among 91 edible plant species. *Asia Pac J Clin Nutr* 17:275-279.

Yarmolinsky L, Huleihel M, Zaccai M and Ben-Shabat S (2012) Potent antiviral flavone glycosides from Ficus benjamina leaves. *Fitoterapia* 83:362-367.

Ye M, Yan Y and Guo Da (2005) Characterization of phenolic compounds in the Chinese herbal drug Tu-Si-Zi by liquid chromatography coupled to electrospray ionization mass spectrometry. *Rapid Communications in Mass Spectrometry* 19:1469-1484.

Yi W, Wei Q, Di G, LIANG J-Y and Yang-Li L (2012) Phenols and flavonoids from the aerial part of Euphorbia hirta. *Chinese journal of natural medicines* 10:40-42.

Yin F, Zhang Y, Yang Z, Cheng Q and Hu L (2006) Triterpene saponins from Gynostemma cardiospermum. *Journal of natural products* 69:1394-1398.

Yoon KD, Jeong DG, Hwang YH, Ryu JM and Kim J (2007) Inhibitors of osteoclast differentiation from Cephalotaxus koreana. *Journal of natural products* 70:2029-2032.

Youn I-S, Han A-R, Roh MS and Seo E-K (2015) Constituents of the leaves of Verbascum blattaria. *Natural product communications* 10:445-446.

Yu X, Zhang T and Wang D (2014) [Chemical Constituents in hypoglycemic active fraction of Celastrus orbiculatus leaf]. *Zhong yao cai= Zhongyaocai= Journal of Chinese medicinal materials* 37:998-1000.

Yuzbasioglu M, Kuruuzum-Uz A, Guvenalp Z, Simon A, Toth G, Harput US, Kazaz C, Bilgili B, Duman H and Saracoglu I (2015) Cytotoxic compounds from endemic Arnebiapurpurea. *Natural product communications* 10:595-596.

Zhang H, Yan L, Zhang Q and Wang Z (2013) [Flavonoids from leaves of Epimedium pubescens]. *Zhongguo Zhong yao za zhi= Zhongguo zhongyao zazhi= China journal of Chinese materia medica* 38:1942-1946.

Zhang J, Huang Z, Huo H-X, Li Y-T, Pang D-R, Zheng J, Zhang Q, Zhao Y-F, Tu P-F and Li J (2015a) Chemical constituents from Lagopsis supina (Steph.) Ik.-Gal. ex Knorr. *Biochemical Systematics and Ecology* 61:424-428.

Zhang J, Li X, Ren L, Fang C and Wang F (2011) [Chemical constituents from Exochorda racemosa]. *Zhongguo Zhong yao za zhi= Zhongguo zhongyao zazhi= China journal of Chinese materia medica* 36:1198-1201.

Zhang L, Yang M, Gao J, Jin S, Wu Z, Wu L and Zhang X (2016) Seasonal variation and gender pattern of phenolic and flavonoid contents in Pistacia chinensis Bunge inflorescences and leaves. *Journal of plant physiology* 191:36-44.

Zhang Q and Sun Z (2014) [Study on Chemical Constituents of Oldenlandia diffusa]. *Zhong yao cai= Zhongyaocai= Journal of Chinese medicinal materials* 37:2216-2218.

Zhang X, Hung TM, Phuong PT, Ngoc TM, Min B-S, Song K-S, Seong YH and Bae K (2006) Anti-inflammatory activity of flavonoids fromPopulus davidiana. *Archives of pharmacal research* 29:1102-1108.

Zhang Y, Chen AY, Li M, Chen C and Yao Q (2008) Ginkgo biloba extract kaempferol inhibits cell proliferation and induces apoptosis in pancreatic cancer cells. *Journal of Surgical Research* 148:17-23.

Zhang Y, Tian H-Y, Tan Y-F, Wong Y-L, Wu HY, Jia J-F, Wang G-E, Gao J-J, Li Y-F and Kurihara H (2015b) Isolation and identification of polyphenols from Marsilea quadrifolia with antioxidant properties in vitro and in vivo. *Natural product research*:1-7.

Zhao L, Peng X, Xia P and Duan W (2014) [Chemical constituents of Tamarix chinensis]. *Zhong yao cai= Zhongyaocai= Journal of Chinese medicinal materials* 37:61-63.

Zhao X, Wang Z, Ma X, Jing W and Liu A (2013) [Chemical constituents from leaves of Nelumbo nucifera]. *Zhongguo Zhong yao za zhi= Zhongguo zhongyao zazhi= China journal of Chinese materia medica* 38:703-708.

Zhao Y, Li H, Ma Q, Li N and Huang S (2011) [Studies on the chemical constituents of the herb of Antenoron filiforme]. *Zhong yao cai= Zhongyaocai= Journal of Chinese medicinal materials* 34:704-707.

Zhen J, Guo Y, Villani T, Carr S, Brendler T, Mumbengegwi DR, Kong A-NT, Simon JE and Wu Q (2015) Phytochemical Analysis and Anti-Inflammatory Activity of the Extracts of the African Medicinal Plant Ximenia caffra. *Journal of analytical methods in chemistry* 2015.

Zhen J, Villani TS, Guo Y, Qi Y, Chin K, Pan M-H, Ho C-T, Simon JE and Wu Q (2016) Phytochemistry, antioxidant capacity, total phenolic content and anti-inflammatory activity of Hibiscus sabdariffa leaves. *Food chemistry* 190:673-680.

Zhou Q, Chen L, Chen Q, Chen H and Dong J (2013) [Chemical constituents of Cudrania cochinchinensis]. *Zhong yao cai= Zhongyaocai= Journal of Chinese medicinal materials* 36:1444-1447.

Zou Q-Y, Wu H-F, Tang Y-L and Chen D-Z (2016) A new labdane diterpene from the rhizomes of Alpinia officinarum. *Natural product research* 30:1-6.

Zou Y-S, Hou A-J, Zhu G-F, Chen Y-F, Sun H-D and Zhao Q-S (2004) Cytotoxic isoprenylated xanthones from Cudrania tricuspidata. *Bioorganic & medicinal chemistry* 12:1947-1953.

Zuo B, Liao Z-X, Xu C and Liu C (2015) Two novel prenylated kaempferol derivatives from fresh bud's fur of Platanus acerifolia and their anti-proliferative activities. *Natural product research*:1-6.

In: Kaempferol ISBN: 978-1-63485-828-1
Editors: T.Garde-Cerdán and A.G. Diago ©2016 Nova Science Publishers, Inc.

Chapter 5

KAEMPFEROL GLYCOSIDES IN *CROCUS*: SOURCES, BIOSYNTHESIS, AND USES

***Natalia Moratalla-López*[1], *Cándida Lorenzo*[1], *Gonzalo L. Alonso*[1] *and Ana M. Sánchez*[2,*]**

[1]Agricultural Chemistry, ETSI Agricultural and Forestry, University of Castilla-La Mancha, University Campus, Albacete, Spain

[2]Unit of Horticulture Research Center and Food Technology of Aragon, CITA. Agrifood IA2 Institute of Aragon (CITA-University of Zaragoza), Zaragoza, Spain

ABSTRACT

The genus *Crocus* (*Iridaceae*) comprises about 160 species occurring in the wild in Europe, the Middle East and North Africa and grown as ornamentals all over the world for their white, yellow, pale-brown, purple to lilac, mauve and blue flowers. Among these species, the cultivated saffron (*Crocus sativus* L.) stands out for the highly valued stigmas of its flowers used as spice, food additive and medicinal drug. Kaempferol glycosides are major flavonoids in the flowers of this genus (70-90% of the total content in the perianth) and are also present in the leaves. In this work, the occurrence of kaempferol and its glycosidic patterns in different *Crocus* species are discussed. Biosynthesis aspects such as

* E-mail: amsanchezgo@aragon.es.

pathways for sequential glycosylation of kaempferol together with the role of kaempferol in the current food, therapeutic or ornamental uses and in other potential uses of the species of the genus *Crocus* are also analysed. Special attention is paid to saffron (*Crocus sativus* L.) and the growing interest of its tepals, which are rich in kaempferol 3-*O*-sophoroside, as a source of antioxidants and active principles.

INTRODUCTION

The genus *Crocus* consists of small corm-bearing perennial species geographically distributed from Western Europe and North-western Africa to Western China, with the centre of species diversity in Asia Minor and on the Balkan Peninsula (Maw, 1886; Mathew, 1982).

Since the comprehensive monographs on the genus published by Maw (1886) and Mathew (1982), the number of recognized species comprised in *Crocus* has steadily increased up to more than 160 species. This ongoing increase is the result of the description of new species and the change of subspecies into species level due to the extensive systematic field investigations of the last decades and the application of molecular methods and phylogenetic studies (Petersen et al., 2008; Harpke et al., 2013, 2014, 2015; Kerndorff et al., 2015a).

Among *Crocus* species, the cultivated saffron (*Crocus sativus* L.) stands out for the highly valued stigmas of its flowers used as spice, food additive and medicinal drug (Figure 1). Moreover, many crocuses are grown as ornamentals all over the world for their white, yellow, pale-brown, purple to lilac, mauve and blue flowers.

The life of a crocus in the majority of the species begins with the germination of the seed and continues by various stages of seedling development in the first year and further development in the following years leading after 3-5 years to the mature plant able to flower (Kerndorff et al., 2015b). However, *C. sativus* is a sterile triploid form and its flowers are unable to produce viable seeds for independent sexual reproduction.

C. sativus has been propagated by corms since at least the year 300 BC (Negbi, 1989) and no wild form is known.

Apart from anthocyanins (delphinidin, petunidin, and malvidin glycosides), kaempferol glycosides are major flavonoids in the flowers of this genus constituting between 70 and 90% of the total content of flavonoids in the perianth (tepals), while quercetin glycosides vary from 5 to 10% and glycosides of dihydrokaempferol, isorhamnetin, myricetin and apigenin are only minor components (Nørbaek, 2002). In stigmas the main flavonoids are kaempferol glucosides, whereas in pollen, kaempferide, a methylated kaempferol, as well as isorhamnetin glycosides are detected (Li and Wu, 2002). Kaempferol and kaempferol glycosides, together with other flavonols (quercetin), flavones (luteolin, tricin, acacetin, apigenin, scutellarein), and their glycosides (Bate-Smidth, 1968; Harborne and Williams, 1984; Williams et al., 1986) are also present in the leaves of *Crocus* species.

In this work, the occurrence of kaempferol, its derivatives, and its glycosidic patterns in different *Crocus* species together with biosynthesis aspects, such as pathways for sequential glycosylation of kaempferol are discussed. After a description of the life-cycle and the function of kaempferol glycosides in *Crocus* plants, the role of kaempferol in the current food, therapeutic or ornamental uses and in other potential uses of the species of the genus *Crocus* are also analysed. Special attention is paid to saffron (*C. sativus*) and the growing interest of its tepals, which are rich in kaempferol 3-*O*-sophoroside, as a source of antioxidants and active principles.

Figure 1. Crocus sativus.

Table 1. ***Crocus*** **species, kaempferol composition, taxonomic assignment within Mathew's system (Mathew, 1982), and distribution of major flavonoids in groups according to Nørbæk et al., (2002)**

Species[a]	**Kaempferol composition**[b]	**Serie**[c]	**Group**[d]
C. abantensis T.Baytop & B.Mathew	tepals: (**1, 3, 5, 6, 11, 15, 18, 23, 24**)a	RETICULATI	II, IV
C. abracteolus Kernd. & Pasche			
C. adamii J. Gay	tepals: (**1, 3, 5, 6, 11, 15, 18, 23, 24**)a	BIFLORI (*C. adamii* sp. complex)	III
C. adamioides Kernd. & Pasche			
C. adanensis T.Baytop & B.Mathew	tepals: (**1, 3, 5, 6, 11, 15, 18, 23, 24**)a	BIFLORI	II
C. aerius Herb.	tepals: (**1, 3, 5, 6, 11, 15, 18, 23, 24**)a	BIFLORI (*C. adamii* sp. complex)	IV
C. akdagensis Kernd. & Pasche		(LYCIOTAURI)	
C. akkayaensis Kernd. & Pasche		(LYCIOTAURI)	
C. alatavicus Regel & Semen.	tepals: (**1, 3, 5, 6, 11, 15, 18, 23, 24**)a	ORIENTALES	IV
C. albocoronatus Kernd.		BIFLORI	
C. aleppicus Baker		ALEPPICI	
C. alexandri Nicic ex Velen.		BIFLORI	
C. almehensis C.D.Brickell & B.Mathew		BIFLORI (*C. adamii* sp. complex)	
C. ancyrensis (Herb.) Maw	tepals: (**1, 3, 5, 6, 11, 15, 18, 23, 24**)a	RETICULATI (RSH)	I, III, IV
C. angustifolius Weston	tepals: (**1, 3, 5, 6, 11, 15, 18, 23, 24**)a	RETICULATI	II, III, IV
C. antalyensis B.Mathew	tepals: (**1, 3, 5, 11**)a b**, (**6, 15, 18, 23, 24**)a	FLAVI	I, II, III
C. antherotes Kernd. & Pasche		(ISAURI)	
C. arizelus Kernd. & Pasche			
C. artvinensis J.Philippow		BIFLORI (*C. adamii* sp. complex)	
C. asumaniae B.Mathew & T.Baytop	tepals: (**1, 3, 5, 6, 11, 15, 18, 23, 24**)a	CROCUS	IV

Species[a]	Kaempferol composition[b]	Serie[c]	Group[d]
C. athous Bornm.			
C. atrospermus Kernd. & Pasche		BIFLORI (LYCIOTAURI)	
C. atticus (Boiss. & Orph.) Orph.	tepals: **1-9** e	RETICULATI	II
C. autranii Albov		KOTSCHYANI	
C. babadagensis Kernd. & Pasche			
C. baytopiorum B.Mathew	tepals: **1-9** e	VERNI (BAYTOPI)	II, III
C. berytius Kernd. & Pasche		(*C. adamii* sp. complex)	
C. beydaglarensis Kernd. & Pasche		(LYCIOTAURI)	
C. bifloriformis Kernd. & Pasche			
C. biflorus Mill.	tepals: (**1, 3, 5, 6, 11, 15, 18, 23, 24**)a	BIFLORI	III, IV
C. boissieri Maw			
C. bolensis Rukšāns*		(SPECIOSI)	
C. boryi J.Gay	tepals: (**1, 3, 5, 6, 11, 15, 18, 23, 24**)a	LAEVIGATI	III
C. boulosii Greuter		ALEPPICI	
C. bowlesianus Kernd. & Pasche		(LYCIOTAURI)	
C. brachyfilus I. Schneid.		(SPECIOSI)	
C. caelestis Kernd. & Pasche		BIFLORI	
C. caeruleus Weston		VERNI	
C. calanthus Kernd. & Pasche		(LYCIOTAURI)	
C. cambessedesii J.Gay		VERSICOLORES	
C. cancellatus Herb.		RETICULATI	
ssp. *cancellatus*		RETICULATI	
ssp. *damascenus* (Herb.) B.Mathew		RETICULATI	
ssp. *lycius* B.Mathew	tepals: (**1, 3, 5, 6, 11, 15, 18, 23, 24**)a	RETICULATI	III, IV
ssp. *mazziaricus* (Herb.) B.Mathew	tepals: (**1, 3, 5, 6, 11, 15, 18, 23, 24**)a	RETICULATI	III, IV
ssp. *pamphylicus* B.Mathew		RETICULATI	
C. candidus E.D.Clarke	tepals: (**1, 3, 5, 6, 11, 15, 18, 23, 24**)a	FLAVI	III
C. caricus Kernd. & Pasche		BIFLORI (ISAURI)	
C. carpetanus Boiss. & Reut.	tepals: (**1, 3, 5, 6, 11, 15, 18, 23, 24**)a	CARPETANI	I, III

Table 1. (Continued)

Species[a]	**Kaempferol composition**[b]	**Serie**[c]	**Group**[d]
C. cartwrightianus Herb.	tepals: (**1, 3, 5, 6, 11, 15, 18, 23, 24**)a	CROCUS	III, IV
C. caspius Fisch. & C.A.Mey. ex Hohen.		BIFLORI	
C. chrysanthus (Herb.) Herb.	tepals: (**1, 3, 5, 6, 11, 18, 23, 24**)a, **12**c**, **15**a d	BIFLORI (ISAURI)	I, III, IV
C. clusii J. Gay		(LONGIFLORI)	II, III, IV
C. coloreus Kernd. & Pasche			
C. concinnus Kernd. & Pasche		(ISAURI)	
C. corsicus Vanucc. ex Maw	tepals: (**1, 3, 6, 11, 15, 18, 23, 24**)a, **5**a c; leaves: **5**c	VERSICOLORES	IV
C. crewei Hook.f.	tepals: (**1, 3, 5, 6, 11, 15, 18, 23, 24**)a	BIFLORI	I, III
C. cvijicii Kosanin	tepals: (**1, 3, 5, 6, 11, 15, 18, 23, 24**)a	RETICULATI	III, IV
C. cyprius Boiss. & Kotschy		BIFLORI	
C. dalmaticus Vis.		RETICULATI	
C. danfordiae Maw	tepals: (**1, 3, 5, 6, 11, 15, 18, 23, 24**)a	BIFLORI	IV
C. danubensis Kernd., Pasche, N. Randjelovic & V. Randjelovic		(RSH)	
C. duplex Weston			
C. elegans Rukšāns		(SPECIOSI)	
C. etruscus Parl.	tepals: (**1, 3, 6, 11, 15, 18, 23, 24**)a, **5**a c, **12**c**; leaves: kaempferol e, **5**c e	VERNI	III, IV
C. *fauseri* Kernd. & Pasche*		BIFLORI (ISAURI)	
C. *fibroannulatus* Kernd. & Pasche		BIFLORI (*C. adamii* sp. complex)	
C. *filis-maculatis* Kernd. & Pasche		(RSH)	
C. *flavus* Weston	tepals: (**1, 3, 5, 6, 11, 15, 18, 23, 24**)a, **5**c	FLAVI	I, IV

Species[a]	Kaempferol composition[b]	Serie[c]	Group[d]
C. *fleischeri* J.Gay	tepals: (**1, 3, 5, 6, 11, 15, 18, 23, 24**)a, **12c****	INTERTEXTI	II, IV
C. *fritschii* Derganc			
C. *gargaricus* Herb.	tepals: (**1, 3, 5, 6, 11, 15, 18, 23, 24**)a	RETICULATI	IV
C. *gilanicus* B.Mathew		KOTSCHYANI	
C. *goulimyi* ssp. *goulimyi*		LONGIFLORI	
C. *graveolens* Boiss. & Reut.	tepals: (**1, 3, 5, 6, 11, 15, 18, 23, 24**)a	FLAVI	III, IV
C. *hadriaticus* Herb.	tepals: (**1, 3, 5, 6, 11, 15, 18, 23, 24**)a	CROCUS	IV
C. *hartmannianus* Holmboe		BIFLORI	
C. *hermoneus* Kotschy ex Maw		RETICULATI	
ssp. hermoneus		RETICULATI	
ssp. palaestinus Feinbrun		RETICULATI	
C. *heuffelianus* Herb.	tepals: (**1, 3, 5, 6, 11, 15, 18, 23, 24**)a	CROCUSIRIS (VERNI)	IV
C. *hittiticus* T.Baytop & B.Mathew	tepals: (**1, 3, 5, 6, 11, 15, 18, 23, 24**)a	RETICULATI (RSH)	III, IV
C. *hybridus* Petrovic			
C. *hyemalis* Boiss. & Blanche		FLAVI	
C. *ibrahimii* Rukšāns		SPECIOSI	
C. *ilgazensis* Rukšāns		SPECIOSI	
C. *ilvensis** Peruzzi & Carta		VERNI	
C. *imperati* Ten.	tepals: (**1, 3, 5, 6, 11, 15, 18, 23, 24**)a	VERSICOLORES	IV
C. *incognitus* Kernd. & Pasche			
C. ionopharynx Kernd. & Pasche		BIFLORI	
C. isauricus Siehe ex Bowles	tepals: (**1, 3, 5, 6, 11, 15, 18, 23, 24**)a	RETICULATI	IV
C. *kangalensis* Kernd. & Pasche		(*C. adamii* sp. complex)	
C. *karamanensis* Kernd. & Pasche		(ISAURI)	
C. *karduchorum* Kotschy ex Maw		KOTSCHYANI	
C. *kartaldagensis* Kernd. & Pasche		(*C. adamii* sp. complex)	

Table 1. (Continued)

Species[a]	**Kaempferol composition**[b]	**Serie**[c]	**Group**[d]
C. katrancensis Kernd. & Pasche		(LYCIOTAURI)	
C. kerndorffiorum Pasche*		BIFLORI	
C. korolkowii Maw & Regel	tepals: **5**c**; leaves: kaempferol e; **5**c	ORIENTALES	I, II
C. kosaninii Pulevic	tepals: (**1**, **3**, **5**, **6**, **11**, **15**, **18**, **23**, **24**)a	VERNI	III, IV
C. kotschyanus K.Koch	tepals: (**1**, **3**, **5**, **6**, **11**, **15**, **18**, **23**, **24**)a	KOTSCHYANI	II, IV
C. laevigatus Bory & Chaub.	tepals: (**1**, **3**, **5**, **6**, **11**, **15**, **18**, **23**, **24**)a, **3** f**; leaves: kaempferol g, **3**f,g	LAEVIGATI	III
C. leichtlinii (Dewar) Bowles	tepals: (**1**, **3**, **5**, **6**, **11**, **15**, **18**, **23**, **24**)a	BIFLORI	III
C. leucostylosus Kernd. & Pasche		BIFLORI	
C. ligusticus Mariotti	tepals: (**1**, **3**, **5**, **6**, **11**, **15**, **18**, **23**, **24**)a	LONGIFLORI	IV
C. longiflorus Raf.	tepals: (**1**, **3**, **5**, **6**, **11**, **15**, **18**, **23**, **24**)a	LONGIFLORI (VERNI)	II, III, IV
C. *lyciotauricus* Kernd. & Pasche		(LYCIOTAURI)	
C. lydius Kernd. & Pasche			
C. malatyensis Kernd. & Pasche			
C. malyi Vis.		VERSICOLORES	
C. marasensis Kernd. & Pasche		(*C. adamii* sp. complex)	
C. mathewii Kernd. & Pasche*		CROCUS	I
C. mawii Kernd. & Pasche			
C. mediotauricus Kernd. & Pasche			
C. melantherus Boiss. & Orph. ex Maw	tepals: (**1**, **3**, **5**, **6**, **11**, **15**, **18**, **23**, **24**)a	BIFLORI	I, IV
C. mersinensis Kernd. & Pasche		(ISAURI)	
C. michelsonii B.Fedtsch.		ORIENTALES	
C. micranthus Boiss		(RSH)	

Species[a]	Kaempferol composition[b]	Serie[c]	Group[d]
C. minimus Redouté	tepals: (**1**, **3**, **6**, **11**, **15**, **18**, **23**, **24**)a, **5**a c, **12**c**	VERSICOLORES	IV
C. minutus Kernd. & Pasche*		BIFLORI	
C. moabiticus Bornm.		CROCUS	
C. multicostatus Kernd. & Pasche			
C. munzurensis Kernd. & Pasche		(*C. adamii* sp. complex)	
C. musagecitii Erol & Yildirim		(*C. adamii* sp. complex)	
C. mysius Kernd. & Pasche			
C. naqabensis Al-Eisawi & Kiswani*		(CROCUS)	
C. neglectus Peruzzi & Carta		LONGIFLORI	
C. nerimaniae Yüzb.*		BIFLORI	
C. nevadensis Amo & Campo		CARPETANI	
C. nivalis Bory & Chaub.	tepals: (**1**, **3**, **5**, **6**, **11**, **15**, **18**, **23**, **24**)a	RETICULATI	II, IV
C. niveus Bowles	tepals: (**1**, **3**, **5**, **6**, **11**, **15**, **18**, **23**, **24**)a	LONGIFLORI	III, IV
C. nubigena Herb.	tepals: (**1**, **3**, **5**, **6**, **11**, **15**, **18**, **23**, **24**)a	BIFLORI	IV
C. nudiflorus Sm.	tepals: (**1**, **3**, **5**, **6**, **11**, **15**, **18**, **23**, **24**)a	LONGIFLORI	II, IV
C. ochroleucus Boiss. & Gaill.	tepals: **5**d	KOTSCHYANI	
C. olivieri J.Gay	tepals: (**1**, **3**, **5**, **6**, **11**, **15**, **18**, **23**, **24**)a	FLAVI	I, III
C. oreocreticus B.L.Burtt	tepals: (**1**, **3**, **5**, **6**, **11**, **15**, **18**, **23**, **24**)a	CROCUS	IV
C. oreogenus Kernd. & Pasche		(LYCIOTAURI)	
C. orphei Karamplianis & Constantin		(RSH)	
C. pallasii Goldb.	tepals: (**1**, **3**, **5**, **6**, **11**, **15**, **18**, **23**, **24**)a	CROCUS	I, III
C. paschei Kerndorff*		FLAVI	
C. paulineae Pasche & Kerndorff			
C. pelistericus Pulevic	tepals: (**1**, **3**, **5**, **6**, **11**, **15**, **18**, **23**, **24**)a	SCARDICI	III, IV
C. pelitensis Kernd. & Pasche		(*C. adamii* sp. complex)	

Table 1. (Continued)

Species[a]	Kaempferol composition[b]	Serie[c]	Group[d]
C. pestalozzae Boiss.		BIFLORI	
C. ponticus Kernd. & Pasche		(*C. adamii* sp. complex)	
C. pseudonubigena B.Mathew	tepals: (**1**, **3**, **5**, **6**, **11**, **15**, **18**, **23**, **24**)a	BIFLORI (*C. adamii* sp. complex)	IV
C. pulchellus Herb.	tepals: (**1**, **3**, **5**, **6**, **11**, **15**, **18**, **23**, **24**)a, **15**d	SPECIOSI	III, IV
C. pulchricolor Herb.	tepals: (**1**, **3**, **5**, **6**, **11**, **15**, **18**, **23**, **24**)a	BIFLORI	IV
C. punctatus B.Mathew	tepals: (**1**, **3**, **5**, **6**, **11**, **15**, **18**, **23**, **24**)a	BIFLORI	IV
C. purpureus Weston		VERNI	
C. rechingeri Kernd. & Pasche		(ISAURI)	
C. reticulatus Steven ex Adams	tepals: (**1**, **3**, **5**, **6**, **11**, **15**, **18**, **23**, **24**)a	RETICULATI (RSH)	III, IV
C. robertianus C.D.Brickell	tepals: (**1**, **3**, **5**, **6**, **11**, **15**, **18**, **23**, **24**)a	RETICULATI	II, III
C. romuleoides Kernd. & Pasche		(*C. adamii* sp. complex)	
C. roopiae Woronow		(*C. adamii* sp. complex)	
C. roseoviolaceus Kernd. & Pasche*		(ISAURI)	
C. rujanensis Randjel. & D.A.Hill*		RETICULATI	
C. sakariensis Rukšāns		SPECIOSI	
C. salurdagensis Kernd. & Pasche		(LYCIOTAURI)	
C. salzmannii J. Gay	tepals: (**1**, **3**, **5**, **6**, **11**, **15**, **18**, **23**, **24**)a, kaempferol f	LONGIFLORI	II, III, IV
C. sativus L.	tepals: **1**a g** h-j, **2**g** i j, **3**a, **4**h i, **5**a g** h-l, **6**a,7p, **9**h, **10**g** h-j, **11**a; **13**i n, **14**h, **15**a, **16**o, **17**i n, **18**a, **19**g**o, **20**g**, **21**g**o, **22**g**, **23-24**a, kaempferol g q** stigmas: **5**n, **8**m**, **13**m**,n, **17**n; leaves: **25-26**r	CROCUS	II, IV

Species[a]	**Kaempferol composition**[b]	**Serie**[c]	**Group**[d]
C. scardicus Kos.		SCARDICI	
C. scharojanii Rupr.		KOTSCHYANI	
C. schneideri Kernd. & Pasche		(*C. adamii* sp. complex)	
C. serotinus Salisb.		LONGIFLORI	
C. sieberi J.Gay	tepals: (**1**, **3**, **5**, **6**, **11**, **15**, **18**, **23**, **24**)a; leaves: kaempferol e	RETICULATI	II, IV
C. sieheanus Barr ex B.L.Burtt		RETICULATI	
C. simavensis Kernd. & Pasche			
C. sivasensis Kernd. & Pasche		(*C. adamii* sp. complex)	
C. speciosus M.Bieb.	tepals: (**1**, **3**, **5**, **11**, **15**, **23**, **24**)a, **6** a b**, **18**a b**, kaempferol f	SPECIOSI	III, IV
C. striatulus Kernd. & Pasche		SPECIOSI	
C. stridii Papan. & Zacharof	tepals: (**1**, **3**, **5**, **6**, **11**, **15**, **18**, **23**, **24**)a	BIFLORI	IV
C. suaveolens Bertol.		VERSICOLORES	
C. sublimis Herb.	tepals: (**1**, **3**, **5**, **6**, **11**, **15**, **18**, **23**, **24**)a	RETICULATI	II, IV
C. suworowianus K.Koch			
C. tahtaliensis Kernd. & Pasche			
C. taseliensis Kernd. & Pasche		(ISAURI)	
C. tauri (Kernd. & Pasche) ex Maw	tepals: (**1**, **3**, **5**, **6**, **11**, **15**, **18**, **23**, **24**)a	BIFLORI (*C. adamii* sp. complex)	I, III
C. tauricus (Trautv.) Puring			
C. thirkeanus K.Koch			
C. thomasii Ten.		CROCUS	
C. thracicus Yüz & Aslan			
C. tommasinianus Herb.	tepals: (**1**, **3**, **5**, **6**, **11**, **15**, **18**, **23**, **24**)a	VERNI	III, IV
C. tournefortii J.Gay		LAEVIGATI	
C. vallicola Herb.	tepals: (**1**, **3**, **5**, **6**, **11**, **15**, **18**, **23**, **24**)a	KOTSCHYANI	III

Table 1. (Continued)

Species[a]	Kaempferol composition[b]	Serie[c]	Group[d]
C. variegatus Hoppe & Hornsch.		(RSH)	
C. veluchensis Herb.	tepals: (**1**, **3**, **5**, **6**, **11**, **15**, **18**, **23**, **24**)a	RETICULATI	III
C. veneris Tapp. ex Poech		ALEPPICI	
C. vernus (L.) Hill	tepals: (**1**, **3**, **5**, **6**, **11**, **15**, **18**, **23**, **24**)a; leaves: kaempferol e	VERNI	III, IV
C. versicolor Ker Gawl.	tepals: (**1**, **3**, **5**, **6**, **11**, **15**, **18**, **23**, **24**)a, **5**c; leaves: kaempferol e, **5**c e	VERSICOLORES	IV
C. vitellinus Wahlenb.	tepals: (**1**, **3**, **5**, **6**, **11**, **15**, **18**, **23**, **24**)a	FLAVI	I, IV
C. wattiorum (B.Mathew) B.Mathew*		BIFLORI	
C. weldenii Hoppe & Fürnr.	tepals: (**1**, **3**, **5**, **6**, **11**, **15**, **18**, **23**, **24**)a	BIFLORI	IV
C. xantholaimos Rukšāns		SPECIOSI	
C. xanthosus Kernd. & Pasche		(LYCIOTAURI)	
C. yaseminiae O. Erol			
C. yataganensis Kernd. & Pasche		BIFLORI	
C. ziyaretensis Kernd. & Pasche		(LYCIOTAURI)	

[a] This list was updated to include new taxa according to GBIF secretariat, 2013; Roskov et al., 2016; Harpke et al., 2014; Kerndorff et al., 2013a, 2013b; Yüzbaşioğlu et al., 2015; Erol et al., 2014, 2015. Species formerly classified as subspecies of *C. biflorus* are indicated by bold lettering. Asterisks (*) indicate species not listed by Mathew (1982) and their probable taxonomic assignment within Mathew's system of Crocus according to Harpke et al., 2013 is given in the corresponding column. Subspecies are shown when data on their flavonoid pattern are available. [b] Kaempferol glycosides are numbered as indicated in Table 2. References are indicated with letters: a = Nørbæk et al., 2002; b = Nørbæk and Kondo, 1999a; c = Harborne and Wiliams, 1984; d = Trapero et al., 2012; e = Williams et al., 1986; f = Price et al., 1938; g = Li et al.,2004; h = Goupy et al., 2013; i = Serrano-Díaz et al., 2014; j = Tuberoso et al., 2016; k = Kubo et al., 1988; l = Sánchez et al., 2011; m = Straubinger et al., 1997; n = Carmona et al., 2007; o = Sánchez et al., 2014; p = Argento et al., 2010; q = Zeka et al., 2015; r = Smolskaite et al., 2011. Double asterisks (**) indicate that in the referenced work, the compound was isolated in quantity for subsequent identification procedures in the corresponding vegetal part of the species. [c] New series classification or circumscription is written in brackets: RSH = Reticulati *sensu* Harpke et al., 2014. [d] According to Nørbæk et al., 2002.

Table 2. Kaempferol glycosides from *Crocus* species

Number	Compound	Mw
1	Kaempferol 3-*O*-β-D-glucoside	448
2	Kaempferol 7-*O*-β-glucoside	448
3	Kaempferol 3-*O*-β-D-(2''-*O*-α-L-rhamnosyl)-glucoside; Synonym: Kaempferol 3-neohesperidoside	594
4	Kaempferol 3-*O*-rutinoside	594
5	Kaempferol 3-*O*-β-D-(2''-*O*-β-D-glucosyl)-glucoside; Synonym: Kaempferol 3-*O*-β-sophoroside	610
6	Kaempferol 3-*O*-α-L-(2''-*O*-β-D-glucosyl)-rhamnoside	594
7	Kaempferol 7-*O*-gentiobioside	610
8	Kaempferol 7-*O*-β-sophoroside	610
9	kaempferol 3-*O*-glucoside-7-*O*-rhamnoside	594
10	Kaempferol 3,7-di-*O*-β-D-glucoside	610
11	Kaempferol 3, 4'-di-*O*-β-D-glucoside	610
12	Kaempferol 3-*O*-β-rutinoside-7-*O*-β-glucoside	756
13	Kaempferol 3-*O*-β-sophoroside-7-*O*-β-glucoside	772
14	Kaempferol 3-*O*-sophoroside-7-*O*-rhamnoside	756
15	Kaempferol 3-*O*-α-L-(2''-*O*-β-D-glucosyl)-rhamnoside-7-*O*-β-D-glucoside	756
16	Kaempferol 3-*O*-glucoside-7-*O*-β-sophoroside	772
17	Kaempferol 3,7,4′-tri-*O*-glucoside	772
18	Kaempferol 3-*O*-α-L-(2','3''-di-*O*-β-D-glucosyl)-rhamnoside	756
19	Kaempferol 3-*O*-β-D-(6-*O*-acetyl)-glucoside	490
20	Kaempferol 3-*O*-β-D-(2''-*O*-β-D-6'''-*O*-acetylglucosyl)-glucoside	652
21	Kaempferol 3-*O*-β-D-(6-*O*-acetyl)-glucoside-7-β-D-glucoside	652
22	Kaempferol 3-*O*-β-D-(2''-*O*-β-D-6-acetylglucosyl)-glucoside-7-*O*- β-D-glucoside	814
23	Kaempferol 3-*O*-α-L-(2''-*O*-β-glucosyl)-rhamnoside-7-*O*-β-(6''''-*O*-acetyl)-glucoside	798
24	Kaempferol 3-*O*-α-L-(2''-*O*-β-D-glucosyl)-rhamnoside-7-*O*-β-D-(6''''-*O*-malonyl)-glucoside	842
25	Kaempferol 6-*O*-β-glucosyl-8-*C*-glucoside	626
26	Kaempferol 3,6-di-*O*-β-glucosyl-8-*C*-glucoside	788

KAEMPFEROL OCCURRENCE AND SOURCES IN *CROCUS*

The genus *Crocus* L. belongs to the order *Asparagales*, family *Iridaceae* and, as already mentioned, comprises more than 160 species (Table 1).

Major flavonoids of many genera of the *Iridaceae* are *C*-glycosylflavones. However, those of *Crocus* species are flavonol-*O*-glycosides.

Kaempferol is mainly present in *Crocus* as monoglycosides, diglycosides or triglycosides. Also, kaempferol acylated glycosides, and *C*-glycosides have been reported in several species of Crocus. Table 2 shows these kaempferol derivatives, and their molecular weight (Mw), where as their occurrence in different *Crocus* species is shown in Table 1.

Kaempferol

The first reports of kaempferol in *Crocus* were those of Price et al., (1938) who isolated free kaempferol from tepals of *C. asturicus* (synonym: *C. salzmannii*), and *C. speciosus*, and Bate-Smith (1968) who found kaempferol to be widespread in hydrolyzed leaf extracts of 49 species (data not shown in Table 1). Regarding kaempferol in hydrolyzed extracts, more recently, Zeka et al., (2015) reported contents extremely high in tepals of *C. sativus* (126 mg g^{-1} dry weight). From the works of Serrano-Díaz et al., (2014) and Tuberoso et al., (2016), where a quantification of kaempferol, and kaempferol glycosides was reported without an specific step of releasing the aglycone from its respective glycosides, much lower contents of kaempferol could be deduced (0.49 and 0.06 mg g^{-1} dry weight, respectively).

Kaempferol Monoglycosides

Structures of kaempferol monoglycosides reported in *Crocus* are shown in Figure 2.

Kaempferol 3-*O*-β-D-glucoside (**1**) was reported as a major flavonol in tepals (Nørbæk et al., 2002). It was isolated from tepals of *C. antalyensis* (Nørbæk and Kondo, 1999a), and *C. sativus* (Li et al., 2004). It was detected in all taxa studied by Nørbæk et al., (2002). In *C. sativus* tepals, contents of 0.29 mg g^{-1} fresh weight, and 1.90 mg g^{-1} dry weight were reported by Goupy et al., (2013). After calculations from data reported by Serrano-Díaz et al., (2014) and Tuberoso et al., (2016), the content of this kaempferol glycoside in saffron floral bio-residues resulted in less than 0.01 mg g^{-1} dry weight for both studies.

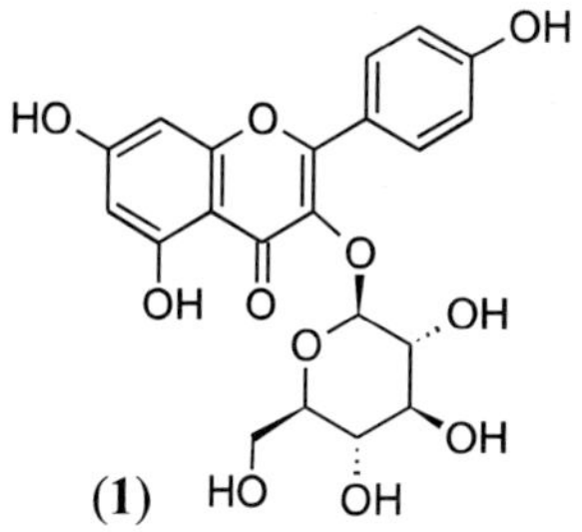

(**1**)

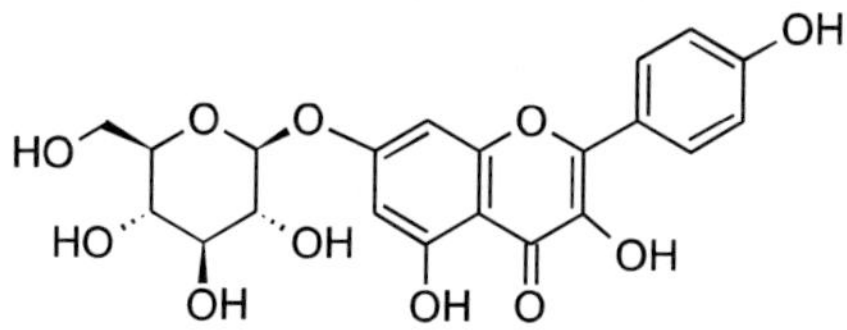

(**2**)

Figure 2. Structures of kaempferol monoglycosides of *Crocus* species.

Kaempferol 7-*O*-β-glucoside (**2**) was isolated from tepals of *C. sativus* (Li et al., 2004). After calculations from data reported by Serrano-Díaz et al., (2014) and Tuberoso et al., (2016), the content of this kaempferol glycoside in saffron floral bio-residues resulted in 0.01, and 0.22 equivalent mg of kaempferol 3-*O*-glucoside g^{-1} dry weight, respectively.

Kaempferol Diglycosides

Figure 3 shows structures of the kaempferol diglycosides reported in *Crocus.*

Kaempferol 3-O-diglycosides

Kaempferol 3-*O*-β-D-(2''-*O*-α-L rhamnosyl)-glucoside (**3**) (synonym: kaempferol 3-*O*-neohesperidoside) was isolated from tepals of *C. antalyensis* (Nørbæk and Kondo, 1999a). It was detected in all taxa studied by Nørbæk et al., (2002).

Kaempferol 3-*O*-β-D-(6''-*O*-α-L-rhamnosyl)-glucoside (**4**) (synonym: kaempferol 3-*O*-rutinoside) was detected in *C. sativus* tepals with contents of 0.04 equivalent mg of kaempferol 3-*O*-glucoside g^{-1} fresh weight and 0.26 equivalent mg of kaempferol 3-*O*-glucoside g^{-1}dry weight by Goupy et al., (2013). After calculations from data reported by Serrano-Díaz et al., (2014), the content of this kaempferol glycoside in saffron floral bio-residues resulted in 0.01 equivalent mg of kaempferol 3-*O*-glucoside g^{-1} dry weight.

3-*O*-diglycosides

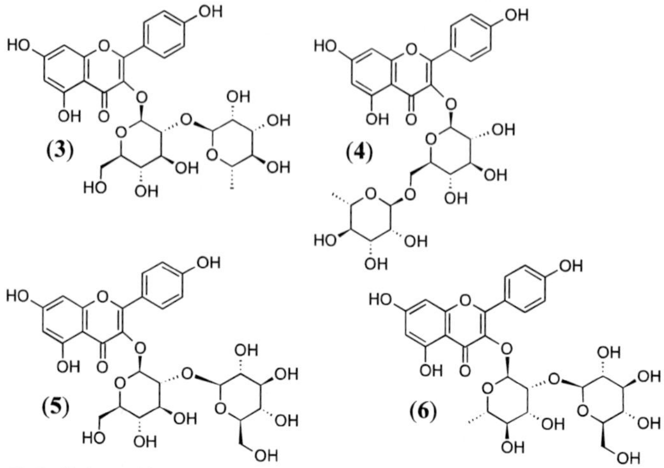

(3) (4) (5) (6)

7-*O*-diglycosides

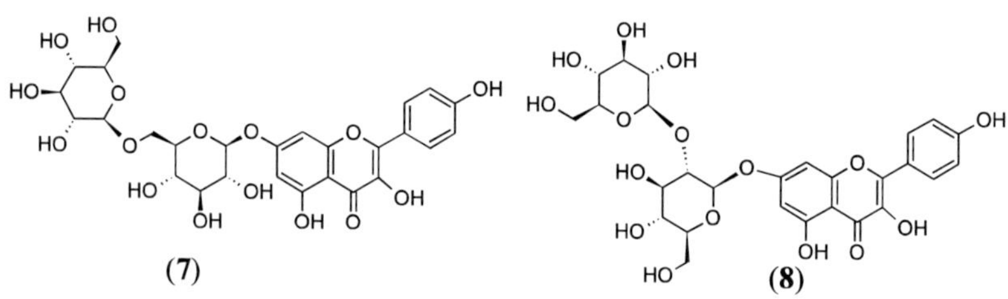

(7) (8)

3,7-di-*O*-glycosides

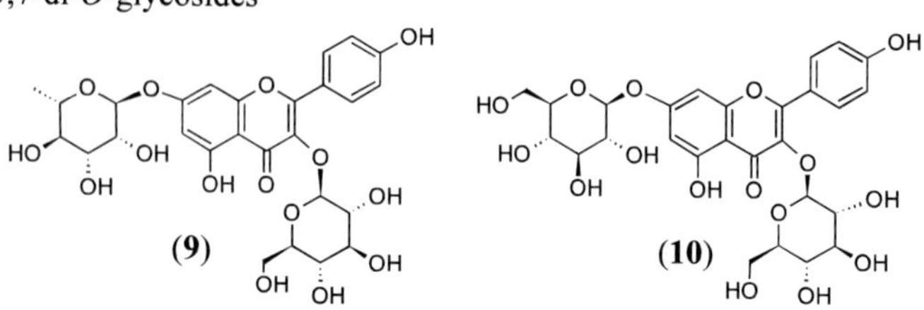

(9) (10)

3,4'-di-*O*-glycosides

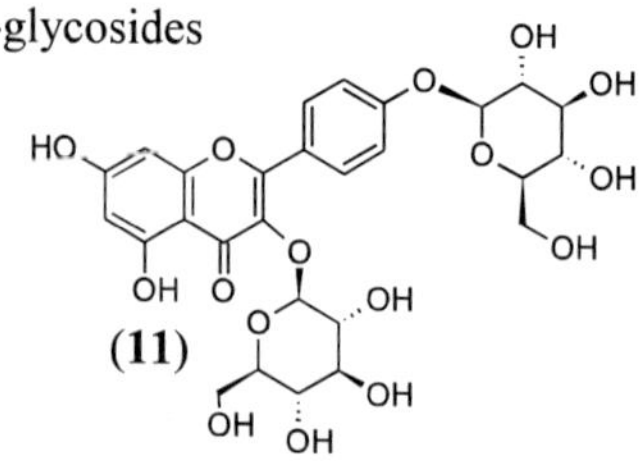

(11)

Figure 3. Structures of kaempferol diglycosides of *Crocus* species.

Kaempferol 3-*O*-β-D-(2''-*O*-β-D-glucosyl)-glucoside (**5**) (synonym: kaempferol 3-*O*-sophoroside) was isolated from tepals of *C. antalyensis* (Nørbæk and Kondo, 1999a), *C. laevigatus*, *C. korolkowii* (Harborne and Williams, 1984), and *C. sativus* (Kubo et al., 1988; Sánchez et al., 2011). It was detected in all taxa studied by Nørbæk et al., (2002). It was also detected in tepals of *C. ochroleucus* (Trapero et al., 2012). In dried stigmas of *C. sativus,* relative contents of kaempferol 3-*O*-sophoroside ranging from 16 to 47% of total flavonoids, and absolute content values ranging from 0.61 to 3.12 equivalent mg of rutin g^{-1} were reported (Carmona et al., 2007). Kaempferol 3-*O*-sophoroside was extracted from saffron floral bio-residues of the production of saffron spice that are mainly made up of tepals, and a extraction yield of 2.3 mg g^{-1} dry weight was reported by Sánchez et al., (2011). According to Goupy et al., (2013), kaempferol 3-*O*-sophoroside represents about 55% of total flavonoids in *C. sativus* tepals, with contents of 1.89 equivalent mg of kaempferol 3-*O*-glucoside g^{-1} fresh weight, and 12.60 equivalent mg of kaempferol 3-*O*-glucoside g^{-1} dry weight. After calculations from data reported by Serrano-Díaz et al., (2014) and Tuberoso et al., (2016), the content of this kaempferol glycoside in saffron floral bio-residues resulted in 0.69, and 2.72 equivalent mg of kaempferol 3-*O*-glucoside g^{-1} dry weight, respectively.

Kaempferol 3-*O*-α L-(2''-*O*-β-D-glucosyl)-rhamnoside (**6**) was isolated from tepals of *C. speciosus* (Nørbæk and Kondo, 1999a). It was detected in all taxa studied by Nørbæk et al., (2002).

Kaempferol 7-O-diglycosides

Kaempferol 7-*O*-gentiobioside (**7**) was found in tepals of *C. sativus* (Argento et al., 2010).

Kaempferol 7 *O*-β-sophoroside (**8**) was first isolated from the dried stigmas of *C. sativus* (Straubinger et al., 1997).

Kaempferol 3,7-di-O-glycosides

Kaempferol 3-*O*-glucoside-7-*O*-rhamnoside (**9**) was detected in tepals of *C. sativus* where contents of 0.01 equivalent mg of kaempferol 3-*O*-glucoside g^{-1} fresh weight, and 0.08 equivalent mg of kaempferol 3-*O*-glucoside g^{-1} dry weight were reported by Goupy et al., (2013).

Kaempferol 3,7-di-*O*-β-D-glucoside (**10**) was isolated from tepals of *C. sativus* (Li et al., 2004). In *C. sativus* tepals, contents of 0.06 equivalent mg of kaempferol 3-*O*-glucoside g^{-1} fresh weight, and 0.42 equivalent mg of kaempferol 3-*O*-glucoside g^{-1} dry weight were reported by Goupy et al., (2013). After calculations from data reported by Serrano-Díaz et al., (2014) and Tuberoso et al., (2016), the content of this kaempferol glycoside in saffron floral bio-residues resulted in 0.01, and 0.02 equivalent mg of kaempferol 3-*O*-glucoside g^{-1} dry weight, respectively.

Kaempferol 3,4'-di-O-glycosides

Kaempferol 3, 4'-di-*O*-β-D-glucoside (**11**) was isolated from tepals of *C. antalyensis* (Nørbæk and Kondo, 1999a). It was detected in all taxa studied by Nørbæk et al., (2002).

Kaempferol Triglycosides

Structures of the kaempferol triglycosides reported in Crocus are shown in Figure 4.

Kaempferol 3-O-diglycoside-7-O-glycosides

Kaempferol 3-*O*-β-rutinoside-7-*O*-β-glucoside (**12**) was isolated from *C. chrysanthus*, *C. fleischeri*, *C. etruscus*, and *C. minimus* (Harborne and Williams, 1984).

Kaempferol 3-*O*-β-sophoroside-7-*O*-β-glucoside (**13**) was first isolated from the dried stigmas of *C. sativus* (Straubinger et al., 1997). In this species it was also detected in tepals (Goupy et al., 2013). In dried stigmas from different geographical origins, contents ranging from 37 to 63% of total flavonoids, and absolute content values ranging from 1.47 to 2.58 of equivalent mg of rutin g^{-1} were reported by Carmona et al., (2007). In tepals, contents of 0.29 equivalent mg of kaempferol 3-*O*-glucoside g^{-1} fresh weight, and 1.90 equivalent mg of kaempferol 3-*O*-glucoside g^{-1} dry weight were reported by Goupy et al., (2013). After calculations from data reported by Serrano-Díaz et al., (2014) and Tuberoso et al., (2016), the content of this kaempferol glycoside in saffron floral bio-residues resulted in 0.11, and 0.03 equivalent mg of kaempferol 3-*O*-glucoside g^{-1} dry weight, respectively.

Kaempferol 3-*O*-sophoroside-7-*O*-rhamnoside (**14**) was reported in *C. sativus* tepals with contents of 0.01 equivalent mg of kaempferol 3-*O*-glucoside g^{-1} fresh weight, and 0.08 equivalent mg of kaempferol 3-*O*-glucoside g^{-1} dry weight by Goupy et al., (2013).

Kaempferol 3-*O*-α-L-(2''-*O*-β-D-glucosyl)-rhamnoside-7-*O*-β-D-glucoside (**15**) was isolated from tepals of *C. chrysanthus-biflorus* cultivars *"Eye-catcher,"* and *"Spring Pearl"* by Nørbæk et al., (1999b). It was detected in all taxa studied by Nørbæk et al., (2002).

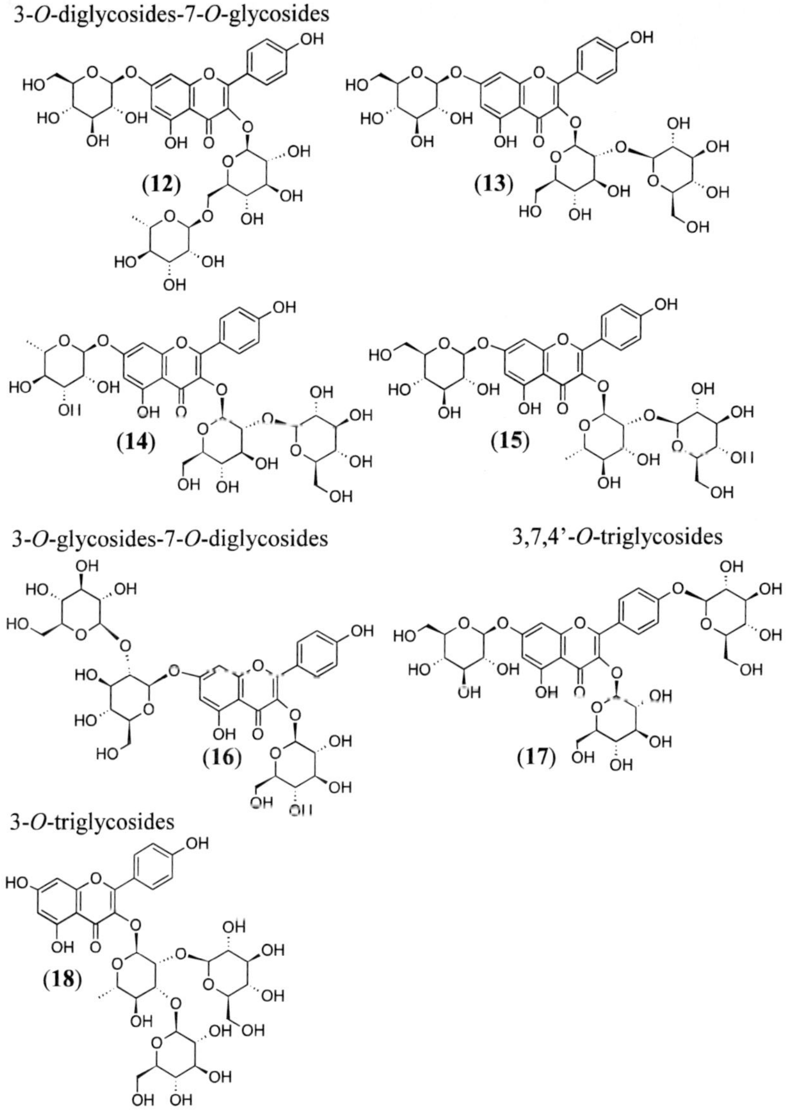

Figure 4. Structures of kaempferol triglycosides of *Crocus* species.

Kaempferol 3-O-glycoside- 7-O-diglycosides

Kaempferol 3-*O*-glucoside-7-*O*-β-sophoroside (**16**) was detected in tepals of *C. sativus* (Sánchez et al., 2014).

Kaempferol 3,7,4'-tri-O-glycosides

Kaempferol 3,7,4′-tri-*O*-glucoside (**17**) was detected in dried stigmas of *C. sativus*, where relative contents ranging from 16 to 22% of total flavonoids, and absolute content values ranging from 0.59 to 1.09 of equivalent mg of rutin g^{-1} were reported (Carmona et al., 2007). It was also detected in the bio-residues of the production of saffron spice with a content value of 0.01 equivalent mg of kaempferol 3-*O*-glucoside g^{-1} dry weight (Serrano-Díaz et al., 2014).

Kaempferol 3-O-triglycosides

Kaempferol 3-*O*-α-L-(2','3''-di-*O*-β-D-glucosyl)-rhamnoside (**18**) was isolated from tepals of *C. speciosus* (Nørbæk and Kondo, 1999a). It was detected in all taxa studied by Nørbæk et al., (2002).

Kaempferol Acylated Glycosides

Structures of the kaempferol acylated glycosides reported in *Crocus* are shown in Figure 5.

Kaempferol 3-*O*-β-D-(6''-*O*-acetyl)-glucoside (**19**), kaempferol 3-*O*-β-D-(2''-*O*-β-D-6'''-*O*-acetylglucosyl)-glucoside (**20**), kaempferol 3-*O*-β-D-(6''-*O*-acetyl)-glucoside-7-β-D-glucoside (**21**), and kaempferol 3-*O*-β-D-(2''-*O*-β-D-6'''-*O*-acetylglucosyl)-glucoside-7-*O*-β-D-glucoside (**22**) were isolated from tepals of *C. sativus* (Li et al., 2004).

Kaempferol 3-*O*-α-L-(2''-*O*-β-glucosyl)-rhamnoside-7-*O*-β-(6''''-*O*-acetyl)-glucoside (**23**), and kaempferol 3-*O*-α-L-(2''-*O*-β-D-glucosyl)-rhamnoside-7-*O*-β-D-(6''''-*O*-malonyl)-glucoside (**24**) were isolated from *C. chrysanthus-biflorus* cultivar *"Spring Pearl"* by Nørbæk et al., (1999b). It was detected in all taxa studied by Nørbæk et al., (2002).

Kaempferol *C*-glycosides

Figure 6 shows the two kaempferol *C*-glycosides were reported in *C. sativus* leaves by Smolskaite et al., (2011): kaempferol 6-*O*-β-glucosyl-8-*C*-glucoside (**25**), and kaempferol 3,6-di-*O*-β-glucosyl-8-*C*-glucoside (**26**).

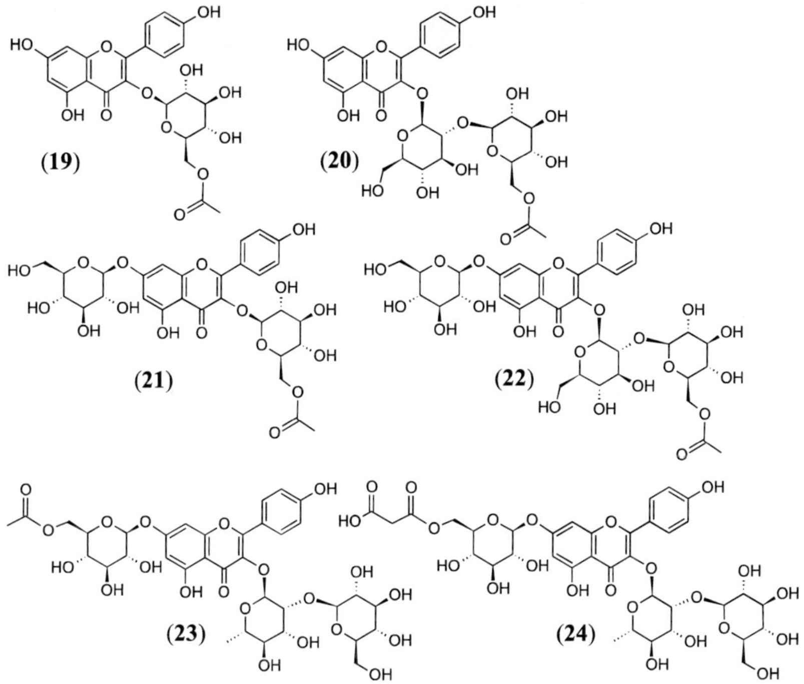

Figure 5. Structures of kaempferol acylated glycosides of *Crocus* species.

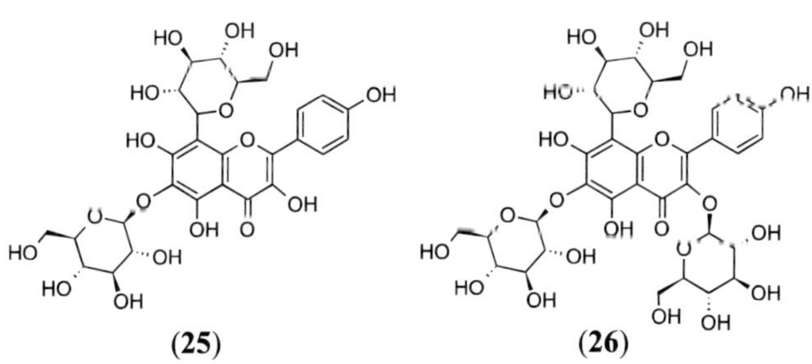

Figure 6. Structures of kaempferol *C*-glycosides of *Crocus* species.

THE GENUS *CROCUS* AND THE CHEMOTAXONOMICAL VALUE OF KAEMPFEROL GLYCOSIDES

In Mathew's classification of *Crocus* (Mathew, 1982) that is the most recent and used one, the 81 recognized species up to then were distributed into two subgenera (subgenus *Crociris* and subgenus *Crocus*). The subgenus *Crociris* consisted only of *C. banaticus* (synonym for *C. heuffelianus* Herb.), and the subgenus *Crocus* was further divided into two sections viz. section *Crocus*, including six series (*Verni*, *Scardici*, *Versicolores*, *Longiflori*, *Kotschyani*, and *Crocus*), and section *Nudiscapus*, which included nine series (*Reticulati*, *Biflori*, *Orientales*, *Flavi*, *Aleppici*, *Carpetani*, *Intertexti*, *Speciosi*, and *Laevigati*) (Table 1). Mathew (1982) also introduced a subspecies concept for many species in different series like in the case of *C. biflorus* Miller and its more than 20 subspecies.

In some species, the various classification characters of Mathew, i.e., distribution pattern, habitat, various morphological traits and cytological data, contributed confusing to the problem of systematic and phylogenetic grouping. Additional chemical characters like floral flavonoids, including kaempferol, were then studied and resulted in valuable characters to supplement the existing classification scheme and distinguish different species (Norbaek et al., 2002).

A number of 70 species and subspecies, 43 cultivars, and six artificial hybrids of *Crocus* were surveyed by Norbaek et al., (2002). These taxa were placed into four chemotypes (I-IV) according to their floral content of flavonoids different to anthocyanins (Table 1). Based on glycosylation patterns, flavonoid group I was defined by a relative amount of 3-*O*-β-neohesperidosides of kaempferol (**5**) and isorhamnetin higher than 20%; flavonoid group II was defined by a relative amount of 3-*O*-β-sophorosides of kaempferol (**3**) and quercetin higher than 20%; flavonoid group III was defined by a relative amount of 3, 4'-di-*O*-β-D-glucosides of kaempferol (**11**), quercetin and isorhamnetin higher than 20%; and group IV was defined by a relative amount of 3-*O*-α-L-(2''-*O*-β-D-glucosyl)-rhamnoside-7-*O*-β-D-glucosides of kaempferol (**15**), quercetin and myricetin; kaempferol 3-*O*-α-L-(2','3''-di-*O*-β-D-glucosyl)-rhamnoside (**18**), kaempferol 3-*O*-α-L-(2''-*O*-β-D-glucosyl)-rhamnoside-7-*O*-β-D-(6''''-*O*-malonyl)-glucoside (**24**), and kaempferol 3-*O*-α-L-(2''-*O*-β-D-glucosyl)-rhamnoside (**6**) higher than 40%. The minor components (7-*O*-β-D-glucosides of dihydro-kaempferol and apigenin) and the major components (3-*O*-β-D-glucosides of kaempferol (**1**) and quercetin) were not included in the grouping because of no chemotaxonomical value. Another kaempferol glycoside that has been proposed for chemotaxonomic purposes is a kaempferol 7-*O*-diglucoside-3-*O*-β-glucoside. This compound and isorhamnetin 3,7-*O*-diglucoside were present in the flowers of *Crocus* species from the series *Crocus*, but absent in *Crocus* species from other series (Ahrazem et al., 2014).

The comparison between the chemotypes defined by Norbaek et al., (2002) and the taxonomical delimitations established by Mathew (1982) revealed that although most chemotypes were represented in several series (Table 1), the chemical data were useful in distinguishing different species. For all series but series *Biflori* the chemical data were very similar for all subspecies or accessions within a species, and chemotypes within a series were more similar than between series. However, for six species, the analyses suggested that they should be further investigated using other methods, to evaluate their relations to other series.

In the last years, molecular methods and phylogenetic studies have brought a completely new understanding of *Crocus*. The methodology has changed and current trend is to compare all taxa which genetically group together phenotypically, morphologically, and geographically in order to arrive at a systematic treatment of *Crocus* that reflects natural relationships among taxa (Kendorff, 2013b).

The phylogenetic analyses clearly indicate that the subspecies concept of *C. biflorus sensu lato* cannot longer be maintained, and that several series of the section *Nudiscapus* stablished by Mathew (1982) either cannot be kept in their original version, e.g., series *Reticulati* (Petersen, 2008; Harpke, 2013, Harpke et al., 2014), or new series have to be introduced, e.g., series *Isauri* (Kerndorff et al., 2014) and series *Lyciotauri* (Kerndorff, 2015a). Also a new series for the *C. adamii* species complex is under preparation by H. Kerndorff (Erol, 2015). Species assigned to these new series are shown in Table 1.

Despite the new understanding of *Crocus*, this genus is still complicated taxonomically and many of the newly described taxa are not assigned to any series (Erol, 2015), as can be seen in Table 1.

BIOSYNTHESIS OF KAEMPFEROL AND ITS GLYCOSIDES IN *CROCUS*

The biosynthesis of flavonoids which have a C6-C3-C6 structure is one of the best known pathways in plant secondary metabolism (Winkel-Shirley, 2001, 2002). The flavonoid biosynthesis pathway represents the convergence of phenylpropanoid and polyketide biosynthetic pathways. The aromatic A-ring is considered to originate from three units of malonyl-CoA, whereas B-ring and the chromane ring are originated from the amino acid phenylalanine that is a product of the shikimate pathway (Figure 7). This system requires two key enzymes: chalcone synthase (CHS, EC 2.3.1.74) and chalcone isomerase (CHI, EC 5.5.1.6). CHS catalyzes the Claisen-like condensation of p-coumaroil-CoA (building block C6-C3) with three molecules of malonyl-CoA (building block C6) resulting in the formation of the naringenin chalcone (C6-C3-C6). Following, by action of the enzyme chalcone isomerase (CHI) occurs the C3 ring closure through Michael-like intramolecular nucleophilic addition to give the flavonoid naringenin, the central precursor flavonoid in this pathway. Then, the kaempferol biosynthesis is completed in two subsequent steps including the hydroxylation of the naringenin at C-3 position by enzyme flavanone-3-dioxygenase (F3H) (EC 1.14.11.9) to give dihydrokaempferol intermediate which finally undergoing dehydrogenation catalyzed by enzyme flavonol synthase (FLS, EC 1.14.11.23) at C2 and C3 position to produce kaempferol. Kaempferol as well as many others flavonoids aglycone are widely distributed in nature owing to the fact that in their biosynthesis there are involved of a set of enzymes fairly common in the plant kingdom (Mena Barreto Silva et al., 2013).

Because the enzymes involved in the biosynthesis of kaempferol are relatively common in the plant kingdom, it is not surprising that this flavonoid is widely distributed in plants. Sugars such as glucose, rhamnose, galactose and rutinose are usually bound to kaempferol to form glycosides. Some glycosides of kaempferol are very common in nature (e.g., kaempferol-3-*O*-glucoside (**1**)), because their biosynthesis only requires additional enzymes that are widespread in the plant kingdom (e.g., flavonol 3-*O*-glucosyltransferase, EC 2.4.1.91). The enzymes involved in the biosynthesis of some other kaempferol glycosides are more restricted in nature and, therefore, these glycosides will only be synthesized by plant species with the genetic information required to code for such enzymes (Calderón-Montaño et al., 2011).

Enzymes that catalyze the glycoside formation, known as glycosyltransferases (UGTs), have been intensively studied. These enzymes transfer the nucleotide-diphosphate-actived sugars (sugar donor) to small molecular weight compounds as substrate acceptor, such as flavonoids (Mena Barreto Silva et al., 2013).

The major enzymatic reactions involved in the flavonoid biosynthesis in *Crocus* were categorised from the presence of flavonoids with corresponding glycosylation patterns by Nørbæk et al., (2002). The enzymes involved in biosynthesis that showed high activity in the taxa corresponding to the different flavonoid groups (Table 1) were: in group I, a flavonoid 3-*O*-glucosyltransferase (FGT, EC 2.4.1.91) a flavonoid 2-*O*-rhamnosyltransferase

(2ORT); in group II, a flavonoid 3-*O*-glucosyltransferase (FGT, EC: 2.4.1.91), a flavonoid 2''-*O*-glucosyltransferase to produce 3-*O*-sophoroside (2OGT[1], EC 2.4.1.239); in group III, a flavonoid 3-*O*-glucosyltransferase (FGT, EC 2.4.1.91), a flavonoid 4'-*O*-glucosyltransferase (F4'OG); and in group IV, a flavonoid 3-*O*-rhamnosyltransferase (FRT), a flavonoid 2''-*O*-glucosyltransferase to produce 3-*O*-(2''-*O*-glucosyl)-rhamnoside (2OGT[2]), a flavonoid 3''-*O*-glucosyltransferase (3OGT), a flavonoid 7-*O*-glucosyltransferase (F7OG), and an acyltransferase (AT) (Figure 8).

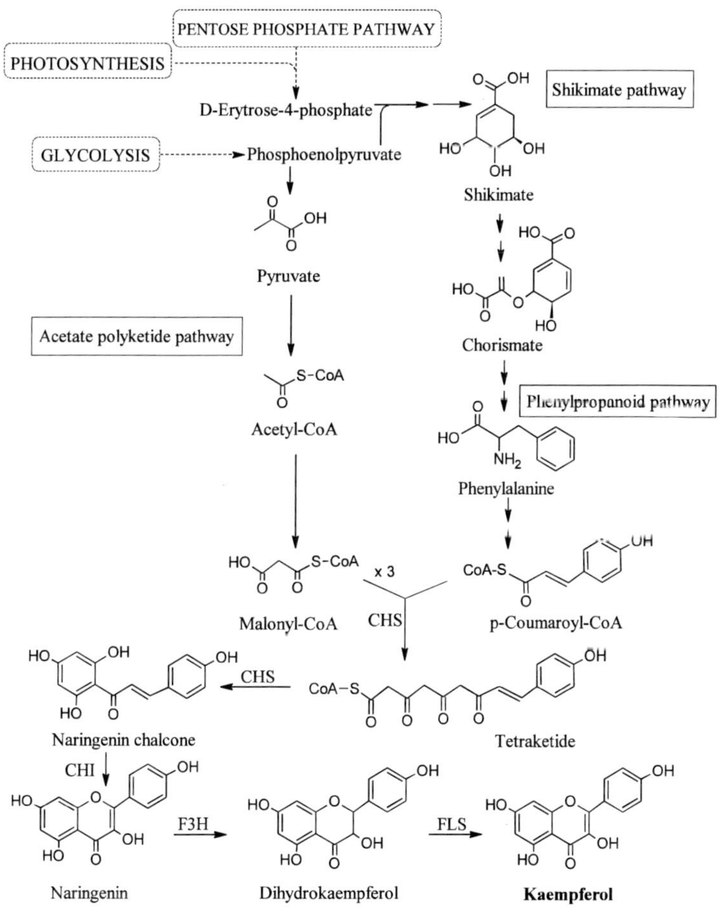

Figure 7. Biosynthesis of kaempferol (adapted from Calderón-Montaño et al., 2011 and Mena Barreto Silva et al., 2013). CHS: chalcone synthase; CHI: chalcone isomerase; F3H: flavanone-3-dioxygenase; FLS: flavonol synthase.

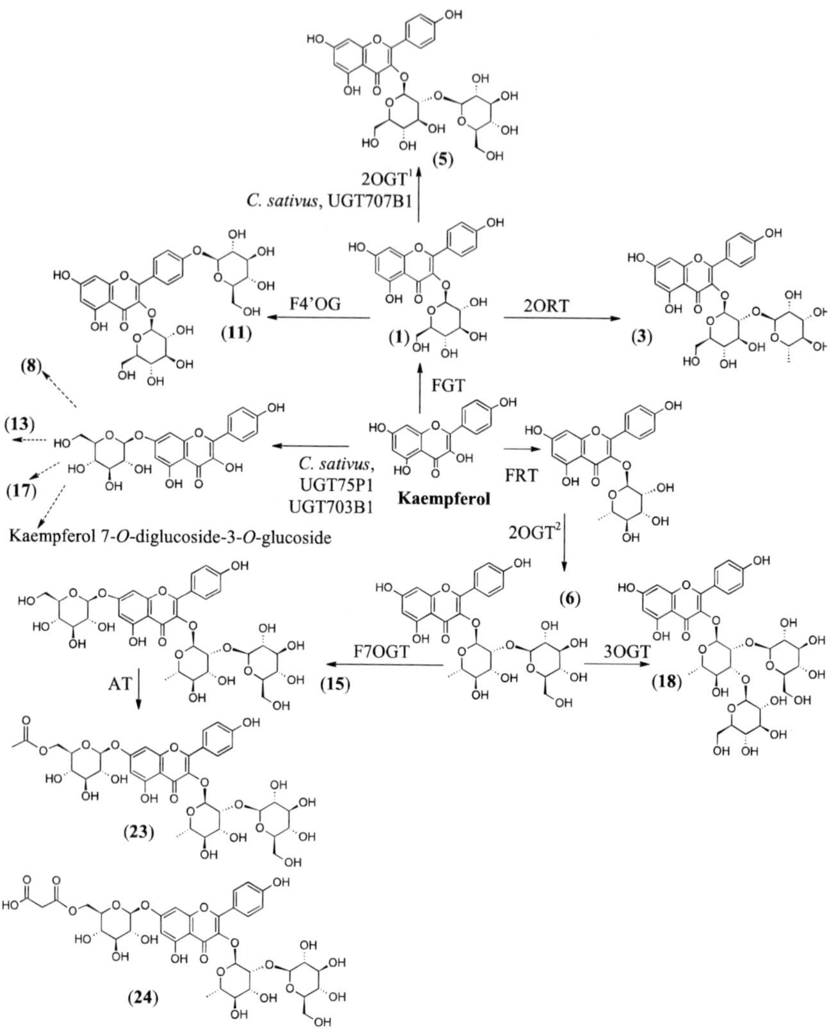

Figure 8. Biosynthesis of some kaempferol glycosides in *Crocus*. FGT: flavonoid 3-*O*-glucosyltransferase, 2ORT: flavonoid 2-*O*-rhamnosyltransferase, 2OGT[1]: flavonoid 3-*O*-glucoside: 2''-*O*-glucosyltransferase, F4'OG: flavonoid 4'-*O*-glucosyltransferase, FRT: flavonoid 3-*O*-rhamnosyltransferase, F7OG: flavonoid 7-*O*-glucosyltransferase, 2OGT[2]: flavonoid 3-*O*-rhamnoside: 2''-*O*-glucosyltransferase, 3OGT: flavonoid 3''-*O*-glucosyltransferase, AT: acyltransferase.

Based on great variation of the ratio: percentages of 3-*O*-β-sophorosides/[percentages of 3-*O*-α-(2''-*O*-β-glucosyl)-rhamnosides + 3-*O*-β-sophorosides] that was calculated for every genotype, Nørbæk et al., (2002) concluded that two different 2-*O*-glucosyltransferases control the formation of flavonol 3-*O*-sophorosides (2OGT[1]) and flavonol 3-*O*-(2''-*O*-glucosyl)-rhamnosides (2OGT[2]), respectively. Similarly, an insignificant variation in the ratio: percentages of kaempferol 3-*O*-α-(2''-*O*-β-glucosyl)-rhamnoside-7-*O*-β-(6'''-*O*-malonyl)-glucoside/[percentages of kaempferol 3-*O*-α-(2''-*O*-β-glucosyl)-rhamnoside-7-*O*-β-(6'''-*O*-malonyl)glucoside + kaempferol 3-*O*-α-(2''-*O*-β-glucosyl)-rhamnoside-7-*O*-β-(6'''-*O*-acetyl)glucoside) gave evidence that the same acyltransferase (AT) was probably catalysing both malonylation and acetylation.

In Figure 8, three different glucosyltransferases that have been characterized in *C. sativus* are also included. The CsGT45, renamed as UGT75P1, was isolated from saffron stigmas. It was characterized as a 7-*O*-glucosyltransferase recognizing kaempferol as a preferential substrate and its expression was associated with the formation of kaempferol-7-*O*-β-sophoroside (**8**), kaempferol-3-*O*-β-sophoroside-7-*O*-β-glucoside (**13**) and kaempferol-3,7,4'-triglucoside (**17**) in certain *Crocus* species (Rubio-Moraga et al., 2009). The enzyme UGT707B1 was suggested to be a flavonol 3-*O*-glucoside: 2''-*O*-glucosyltransferase involved in the biosynthesis of flavonol-3-*O*-sophorosides, in particular of kaempferol-3-*O*-ß-sophoroside (**5**) (Trapero et al., 2012). The last glucosyltransferase characterized in saffron, UGT703B1, showed activity and high regiospecificity toward kaempferol by preferably glucosylating the C7 hydroxyl group. Its expression was correlated with the presence of kaempferol-7-*O*-diglucoside-3-*O*-β-glucoside (Ahrazem et al., 2014).

Life-Cycle and Function of Kaempferol Glycosides in Crocus Plants

General aspects of the Life-Cycle of a Crocus

Following is the description of the life-cycle of a crocus and its changes through time according to Kerndorff et al., (2015b).

Life-Cycle during the First Season

The autumn-flowering crocuses generally germinate from September to November; the vernal species from mid-winter to spring. Crocus seeds have prolonged germination ability and many seeds only germinate after another year or even more years dormant in the ground. In the first year, the crocus produces only one leaf to create a corm that goes into dormancy when hot and dry weather conditions arrive. At that time, all parts of the young crocus die away except the corm. During dormancy stage, corm apparently shows neither external growth, nor morphological change, although there do exist internal physiological and morphogenetic changes. This corm normally leaves dormancy for a new life-cycle when temperature in the soil drops.

Life-Cycle during the Second Season

After the first season, two cases of dormancies have to be distinguished:

a) High-mountain crocuses seem to have a prolonged dormancy and after summer nothing of the "reawakened" crocus appears above ground until the following spring when conditions get favourable again. Meanwhile, predominantly in autumn, the crocus produces new roots and shoots of new leaves to be ready for the next season. If conditions allow, on warm days in spring when snow melts away and roots get sufficient water and nutrients, the young crocus pushes up two or more leaves above ground.

b) Lowland crocuses have a shorter dormancy with leaves appearing above ground in autumn and winter when the soil is wet enough.

Life-Cycle from the Second Season until Corm Maturity

The behaviours of growing crocuses in high-mountain and lowland scenarios are repeated season for season until the crocus reaches its maturity and is able to flower (normally after 3-5 years). From the second season onwards the new corm is built on top of the old one by consuming the old one totally until next dormancy. This means that from the second season onwards the whole plant renews itself every year.

Life-Cycle for Mature Crocuses

Several cases can be distinguished depending on the flowering time:

a) High mountain crocuses flowering in autumn. They flower without leaves above ground. Fertilized flowers wither quickly away, and the capsules containing the growing seeds develop slowly underground. In the following spring when conditions get dry and warm again, leaves and pedicels grow quickly. The capsule is then pushed significantly above ground by the pedicel and, once ripe, it sets the seeds free. There are, however, exceptions of this behaviour which ripen their capsules underground. After seed dispersal and corm-ripeness, the leaves retract and the crocus goes to dormancy again.

b) High mountain crocuses flowering in spring. After a short summer, with cold days coming up, the crocus starts to produce leaf- and flower-shoots but all stay underground. With increasing wetness, roots also start to grow. At the end of winter or when conditions are favourable for plant-life again, the crocus is able to develop flowers within hours, accompanied by more or less strong leaf-development in the following weeks. In case of fertilization, the flowers quickly shrivel away and capsule and seeds grow fast in the short summer. The pedicel pushes the capsule above ground so that the ripe seeds can be dispersed in dry weather. Before the crocus goes into dormancy again the new corm matures on top of the old one.
c) Lowland crocuses flowering in autumn. After dormancy the crocus starts to flower from October to December by pushing up the "ready-to-go" flower-shoots. Most of the crocuses in these areas show a parallel leaf-development, although in some species the leaf-development above ground occurs even before the flower is built.
d) Lowland crocuses flowering in spring. They generally develop their leaves above ground before or with the flowers. Their life-cycle is, therefore, not very different from the autumn-flowering one, except for their flowering-time and, consequently, the necessity to ripen the seeds in a shorter period, since the ripening of the seed and dehiscence of the capsule takes place in June or July.

For drawings, pictures and more details about life-cycle, morphological and phenotypical characteristics of *Crocus* the reader is referred to Kerndorff et al., (2015b).

Life-Cycle of Cultivated Saffron (*C. sativus*)

López-Córcoles et al., (2015) described the annual cycle and the principal growth stages of saffron in Albacete, Spain. Its life-cycle starts with the dormant corm. After dormancy, a root plate is formed at the base of the corm, and one or more shoots subsequently emerge with the leaves and flowers wrapped by the cataphylls at the end of October. The above-ground plant growth derived from sprouting is due to available reserves in the corm. Flowering occurs from the second fortnight of October to the first fortnight of November (ITAP, 2013), and previously to leaf appearance or simultaneously. Later on, leaves and roots keep on growing along the period of vegetative activity. The base of shoots begins to swell forming the daughter corms while the mother corm depletes its reserves. In spring, as diurnal temperatures increase and soil moisture decrease, the mother corm roots die and break away, photoassimilates move from the leaves to the corms and the leaves start to senescence from the apex to the base; the daughter corms are fully developed and ready for latency in June.

Function of Kaempferol Glycosides in *Crocus* Plants

In plants, flavonoids have many diverse functions, such as defence, UV protection, allelopathy, regulation of reactive oxygen species, influence on pollen viability, flower colouring, attraction of pollinators and auxin transport inhibition (Buer et al., 2010). Evidence of these functions and the presence of kaempferol derivatives on *Crocus* species have been mainly reported dealing with flower colouring, auxin transport, floral morphogenesis and pollination.

In tepals of *Crocus* species, flavonoids, and in particular kaempferol glycosides, could directly contribute to colour formation in the purple, lilac, mauve, and blue flowers by copigmentation with anthocyanins where interactions may depend on the sugar units. As copigments, flavonols and flavones do not by themselves significantly contribute to the colour, but they cause a bathochromic wavelength shift in the visible λ_{max} of natural anthocyanins, which makes them appear bluer and increase their absorptivity (Asen et al., 1972, 1975; Scheffeldt and Hrazdina, 1978; Kim and Fujieda, 1991; Takemura et al., 2005).

On the other hand, flavonols and flavones, although having a weak yellow colour as pure substances, do not appear to contribute to yellow flower colour, since their glycosides occur in just as high a concentration in white flowered species as in yellow ones (Harborne and Williams, 1984). Yellow colour of the flowers of some *Crocus* species is mainly carotenoid-based due to crocetin glycosides (Harborne and Williams, 1984; Rubio-Moraga et al., 2013).

It is known that auxins participate in the elongation of the floral tube in *Crocus* (Stark, 1982). In *C. sativus* the accumulation of a specific set of flavonols may influence the transport of the plant hormone auxin and could be responsible for the morphological characteristics of the stigma. Besides, Trapero et al., (2012) reported that distinct parts of the female organs of saffron differed in their flavonol contents, and this distribution could have a direct role in the control of the auxin gradient. These authors suggested the involvement of flavonol 3-*O*-sophorosides, including kaempferol 3-*O*-sophoroside and quercetin 3-*O*-sophoroside, in auxin transport.

Furthermore, not only could the accumulation of flavonols be associated with floral morphogenesis, but also with pollination (Mo et al., 1992). In stigmas of *C. sativus*, the highest accumulation of kaempferol sophorosides has been observed at the time of the anthesis (Rubio-Moraga et al., 2009), like in other plant species (Beliaeva and Evdokimova, 2004).

USES OF *CROCUS*

Food Use of *Crocus*

Food Additive: Saffron Spice

Main concern in the *Crocus* genus is related to the species *C. sativus*, commercially cultivated to produce saffron spice, the world's most expensive spice by weight with a long history of use in traditional medicines of many cultures. Saffron spice consists of the dried stigmas of *C. sativus* blossoms. *C. sativus* is cultivated almost exclusively for its stigmas, which after extremely laborious manual harvesting and processing give rise to this spice. Saffron spice is highly valued both in cookery and in the food industry due to its colouring properties, pleasant bitter taste and characteristic aroma.

Nowadays the main producer country of saffron spice is Iran. According to Iranian Ministry of Agriculture, in the last harvest Iran produced about 94% of world production, with more than 280 tons (The Iran Project, 2015). India, Greece, Morocco, Spain, Italy and Turkey are also involved in the production of saffron. Small productions are also found in countries like France, Switzerland, China, Afghanistan, Azerbaijan, Japan, Tasmania, New Zealand, Argentina, Mexico, the United States or Portugal.

In Europe, acreage of saffron crop in the last century has declined dramatically, disappearing in some regions. Greece, Spain and Italy are the largest European producers, whereas Switzerland retains a small annual production (1-4 kg) marketed under the Protected Designation of Origin (PDO) "Munder Safran." Production in Greece is concentrated in the region of Kozani, whose PDO is "Krokos Kozanis" (OJ, 1999). In Spain, the acreage of saffron crop was 166 ha with a production of 1,918 kg in 2013 (MAGRAMA, 2015). Castilla-La Mancha is the autonomous region that produces most of saffron in Spain. Part of this production is marketed under the PDO "Azafrán de La Mancha" (OJ, 2001), whose outstanding quality is recognized internationally. During 2014, Spain exported 133,825 kg of saffron to 71 countries, being the main destinations, England, France and Belgium (Comtrade, 2015). In Italy, production of saffron is mainly located in Cerdeña (L´Aquila), some areas of the Toscana and Sicily, with three PDOs: "Zafferano dell´Aquila," "Zafferano di San Gimignano" (OJ, 2005) and "Zafferano di Sardegna" (OJ, 2009).

Saffron spice is part of some of the best known traditional dishes from several Mediterranean countries: *Risotto alla Milanense* in Italy, *Bouillabaisse* in France and *Paella Valenciana* in Spain. It is also an indispensable ingredient in various pastries and breads, many of them commemorative of a lay or religious festival, such as the German *Gugelhupf*, Saint Lucia buns in Sweden, buns and breads in Cornwall (England), Christmas bread in Estonia, sweets on the Greek islands, rice pudding in Iranian Shiite celebrations, or the Jewish Sabbath *Challah* bread. In industry, saffron spice is also used to produce several liqueurs and vermouths, sauces and milk products (Carmona et al., 2006a).

The quality of saffron in international commercial agreements is determined according to the ISO 3632 standard, which classifies it into three categories depending upon their physical and chemical characteristics. Colour is the most important quality characteristic and colouring strength values ($E_{1\,cm}^{1\,\%}$ 440 nm) of its aqueous extracts are critical for the commercial value of the spice. The red color of saffron spice and its colouring capacity to give yellowish-red hues are due to crocetin esters, a group of water-soluble carotenoids that derive from crocetin ($C_{20}H_{24}O_4$, 8,8′-diapo-Ψ,Ψ′-carotenedioic acid) where glucose, gentiobiose, neapolitanose or triglucose are the sugar moieties and where *trans* or *cis* configuration is found. These apocarotenoids are also known as crocins (Carmona et al., 2006b). Picrocrocin (4-(β-D-glucopyranosyloxy)-2,6,6-trimethyl-1-cyclohexene-1-carboxaldehyde) is thought to be the foremost contributor to the bitter taste of saffron, although kaempferol glycosides, picrocrocin-related compounds, and amino acids with this organoleptic property have been characterized in saffron spice (Straubinger et al., 1998; Carmona et al., 2007; Del Campo et al., 2009). Safranal is the major compound in the saffron volatile fraction contributing to its aroma. Picrocrocin is converted to safranal either by a two-step enzymatic/dehydration process involving the intermediate 4-hydroxy-2,6,6-

trimethyl-1-cyclohexen-1-carboxaldehyde (HTCC) or directly by thermal degradation. The importance of the saffron quality control makes new and more accurate methods for determination of the crocetin esters, picrocrocin and safranal to be continuously developed (García-Rodríguez et al., 2014).

It is known that crocins are found, not only in the stigmas of *C. sativus*, but also in the stigmas, and tepals of yellow *Crocus* species, like *C. ancyrensis*, *C. angustifolius*, *C. chrysanthus*, *C. olivieri*, *C. korolkowii*, *C. vitellinus*, and *C. sieberi* (Rubio-Moraga et al., 2013). Several references on the collection of the stigmas of several wild species of *Crocus* are found in Maw's monograph (1886). These references reported that in Syria the stigmas of *C. cancellatus* were collected, dried in the sun or pressed into small tablets, and sold in the Bazaars. It was also reported that stigmas were collected from *C. ancyrensis* in Asia Minor; from *C. longiflorus* in Sicily; from *C. thomasii* in South Italy where they were used for the flavouring of dishes; and from *C. cartwrightianus* on the higher parts of the Island of Andros where a pigment was prepared. Maw indicated that the stigmas of the *Crocus* were simply dried as they were gathered, or they were pressed in the process of drying into compact cakes.

Food

Maw's monograph (1886) reported the largely use of small wild corms of several species of Crocus as food in Syria and Asia Minor, where they had a regular commercial value. *C. cancellatus* was so abundant in the Lebanon and the neighbourhood of Damascus that the corms were collected for food. Corms of *C. ancyrensis* were also collected for food at Sivas, in Asia Minor. Corms of *C. gaillardotii* (synonym for *C. aleppicus* Baker) were so common in the neighbourhood of Damascus that they were sold as food in the markers, together with the corms of *C. edulis* (synonym for *C. cancellatus subspecie damascenus* (Herb.) B. Mathew). Looking at their nutritious composition, more than half the weight consisting of sugar and starch. Maw suggested that their culture and use as an article of food might be worth consideration.

Flowers of *C. sativus* have been traditionally consumed as sweets in Sardinia (Italy) (Serrano-Díaz, 2013). In the last years, the possibilities of mechanization and modernization of saffron crop have given rise to a growing interest on the development of new food products from *C. sativus*, beyond saffron spice, such as whole flowers without separating stigmas, or other floral parts different to stigmas. About 68 kg of flowers are needed to produce 1 kg of saffron spice, while 63 kg of bio-residues composed of tepals, stamens and styles are generated (Serrano-Díaz et al., 2013). The studies focused on the nutritional and antioxidant value of the flowers and the floral bio-residues of the saffron spice production are initial contributions to the development of new food products from the flowers of *C. sativus* (Sánchez-Vioque et al., 2012; Serrano-Díaz et al., 2012; Goupy et al., 2013; Serrano-Díaz et al., 2013).

According to Serrano-Díaz et al., (2013), whole flowers and floral bio-residues of *C. sativus* provide low energy and lipids, and high carbohydrate content. Whole flowers have high ash (7.39 mg/100 g), protein (10.07 mg/100g), and available carbohydrates (61.2 mg/100 g), and are low in lipids (3.16 mg/100 g). The contents of Ca, K, Mg, P, and especially of Fe are high. Stamens are the flower parts with the highest ash (11.43 mg/100 g), protein (24.05 mg/100 g), lipids (10.73 mg/100 g), total dietary fiber (32.2 mg/100 g) and insoluble dietary fiber (21.1 mg/100 g) contents, and also the lowest content of available carbohydrates (33.8 mg/100 g) and total sugars (4.3 mg/100 g). On the other hand, the whole flowers, each of its parts and the floral bio-residues are rich in dietary fiber. The ratios insoluble/soluble dietary fiber of floral bio-residues (1.2), stamens (1.9) and stigmas (1.3) are suitable as a balanced source of both types of dietary fiber, since they are in the range of 1.0-2.3 necessary to obtain the beneficial physiological effects associated with both dietary fiber fractions (Spiller, 1986).

Animal Food

Leaves of *C. sativus* are considered a good source of animal feed. The leaves can be used directly as grazing or after harvesting as hay for animal feed. In Spain, it has been reported that these leaves, which could acquire about 40 cm of length and appeared in winter, could serve as feed for animals, especially for females, since they improved and increased their milk secretions (Alarcón and Sánchez, 1968). In Khorasan (Iran), it was considered the possibility of using leaves as a forage source with intermediate quality and digestibility for sheep and goats (Valizadeh, 1988). Valizadeh (1989) reported that more than 75% of Iranian farmers used saffron leaves as hay for animal feed and 14% prefer direct grazing.

Therapeutic Use of Crocus

There are evidences that kaempferol and its glycosides induce cell death in a variety of cancer cell lines, derived from different tissues (Luo et al., 2010), reduce the resistance of cancer cells to anti-cancer drugs such as viniblastine and paclitaxel (Limtrakul et al., 2005), have antityrosinase (Kubo and Kinst-Hori, 1999) and antioxidant activities (Li et al., 2004), and may be a valuable agent in the treatment of depression (Hosseinzadeh and Nassiri-Asl, 2013).

Kaempferol 3-*O*-sophoroside has anti-inflammatory effects in human endothelial cells and analgesic activity (Kim et al., 2012).

The consumption of these flavonols appears to reduce the risk of pancreatic and lung cancer (Ute et al., 2007), can slow skin aging by inhibiting a variety of enzymes that degrade the extracellular matrix, such as elastases, collagenases and hyaluronidases (Hadizadeh et al., 2003).

From antiquity, saffron spice has been valued worldwide for its traditional medicinal use and for a range of potential clinical and pharmaceutical uses. Saffron is beneficial in the treatment of several digestive disorders, especially valuable in flatulent colic and in the treatment of urinary problems. It is used in the fevers, melancholia and enlargement of the liver and spleen. It is employed in medicines that reduce inflammation and as a treatment for pertussis and asthma. It is also used as a strengthening agent for the heart and as a cooling agent for the brain. It is useful in promoting and regulating menstrual periods, and it soothes lumbar pains, which accompany menstruation (Licón et al., 2010; Ulbricht et al., 2011; Wani et al., 2011). *In vitro* cell toxicity tests prove that saffron stigma extracts inhibit the growth and synthesis of nucleic acids in tumor cells, while normal cells are less susceptible and even completely unsusceptible (Abdullaev and Frenkel, 1992). Thus, the evidences indicate that saffron possesses anticancer property against a wide spectrum of tumors, such as leukemia, prostate, ovarian carcinoma and colon adenocarcinoma (Abdullaev and Espinosa-Aguirre, 2004; Zhang et al., 2013; Festuccia et al., 2014). Crocetin is considered a safer alternative to treat *all-trans* retinoic acids sensitive cancers in women of childbearing age (Martina et al., 2002). On the basis of *in vitro*, animal, and human evidence, constituents of saffron may exert antioxidant effects (Kanakis et al., 2007; Saleem et al., 2006; Verma and Bordia, 1998).

It has been proposed that saffron is effective against arteriosclerosis by reducing cholesterol levels in the blood, and it offers protective effects against cardiovascular disease, diabetes, Parkinson's disease, and apoptosis (Bhargava, 2011). The validity of saffron extracts in the treatment of mild to moderate depression has also been confirmed (Noorbalaa et al., 2005; Lopresti and Drummond, 2014).

Saffron floral bio-residues are nowadays receiving a lot of attention because of their biomedical properties, such as cytotoxic effect against tumor cell lines, antifungal and antioxidant (Zheng et al., 2011). It is known that saffron flowers extracts have biological activities, showing antityrosinase activity (Li et al., 2004; Sariri et al., 2011), antioxidant and antiradical activities (Serrano-Díaz et al., 2012; Tuberoso et al., 2016), and metal chelating properties (Sánchez-Vioque et al., 2012). Besides, they have other kind of properties like: anti-inflammatory and antinociceptive activity (Hosseinzadeh and Younesi, 2002), cytotoxic and antifungal activity against tumor cell lines (Zheng et al., 2011), arterial pressure reducer (Fatehi et al., 2003) and they have some benefits in the treatment of mild-to-moderate depression (Moshiri et al., 2006).

Ornamental Use of *Crocus*

Many *Crocus* species are highly appreciated as garden plants for their showy and colourful flowers. Besides, the recorded tolerance to summer drought and winter cold makes wild species interesting as ornamental in areas with severe climatic conditions. Wild species can also be used as sources of useful genes in breeding programmes of the cultivated species (Biodiversity International et al., 2015).

In 2011, the world ornamental plant area was 209,100 ha. Europe cultivated 43% of this area, and China was the second major producer with 15% of the cultivated area. The Netherlands is the largest producer of flower and ornamental plants in Europe, with 21.4% of crop area in Europe. It is followed by Germany (13.9%) and Spain (12.3%) (Eurostat, 2011). Statistics for spring flowering crops in The Netherlands during period 2014-2015 reported a bulbous plant total area of 14,590.6 ha, mainly cultivated with *Tulipa*, *Lilium*, *Narcisus*, *Gladiolus*, *Hyacinthus*, *Iris* and *Crocus* species. Of this total crop area, 407.6 ha corresponded to *Crocus* (BKD, 2016).

The most economically important *Crocus* species as ornamentals are *C. vernus*, *C. biflorus*, *C. chrysanthus*, *C. serotinus*, *C. speciosus*, *C. tommmasinianus* and *C. sieberi*. Among them, *C. vernus*, *C. biflorus*, and *C. chrysanthus* have been hybridized extensively to produce important ornamentals (Wiersema and León, 2013).

Acknowledgment

N. Moratalla-López wishes to thank the Universidad de Castilla-La Mancha for the predoctoral contract PREDUCLM15-35.

References

Abdullaev FI and Frenkel GD (1992) The effect of saffron on intracellular DNA, RNA and protein synthesis in malignant and nonmalignant human cells. *Biofactors* 4:38-41.

Abdullaev FI and Espinosa-Aguirre JJ (2004) Biomedical properties of saffron and its potential use in cancer therapy and chemoprevention trials. *Cancer Detection and Prevention* 28:426-432.

Ahrazem O, Rubio-Moraga A, Trapero A and Gómez-Gómez ML (2014) Genomic organization of a UPD-glucosyltransferase gene determines differential accumulation of specific flavonoid glucosides in tepals. *Plant Cell, Tissue and Organ Culture* 119:227-245.

Alarcón J and Sánchez A (1968) El azafrán. Hojas divulgadoras del Ministerio de Agricultura 18:18. Madrid.

Argento S, Branca F, Siracusa L, Strano T, Napoli EM and Ruberto G (2010) Re-evaluation of saffron floral wastes: analysis of saffron flowers defatted hydro-alcoholic extracts. *Acta Horticulturae* 850:251-260.

Asen S, Stewart RN and Norris K. H (1972) Co-pigmentation of anthocyanins in plant tissues and its effect on color. *Phytochemistry* 11:1139-1144.

Asen S, Stewart RN and Norris KH (1975) Anthocyanin, flavonol copigments and pH responsible for larkspur flower color. *Phytochemistry* 14:2677-2682.

Bate-Smith EC (1968) The phenolic constituents of plants and their taxonomic significance. *Journal of the Linnean Society of London, Botany* 60:325-356.

Beliaeva RG and Evdokimova LI (2004) Variability of flavonol contents during floral morphogenesis in *Papaver somniferum* L. *Ontogenez* 35:16-22.

Bhargava VK (2011) Medicinal uses and pharmacological properties of *Crocus sativus* Linn. (saffron). *International Journal of Pharmacy and Pharmaceutical Sciences* 3:22-26.

Bioversity International, UPV, UCLM and JCCM (2015) Descriptors for *Crocus* (*Crocus* spp.) Bioversity International, Rome, Italy; Universitat Politecnica de Valencia, Valencia, Spain; Universidad de Castilla-La Mancha, Ciudad Real, Spain; and Junta de Comunidades de Castilla-La Mancha, Cuenca, Spain.

BKD (2016) Voorlopige statistiek voorjaarsbloeiers 2014-2015. http://www.bkd.eu/over-de-bkd [15th April 2016].

Buer CS, Imin N and Djordjevic MA (2010) Flavonoids: new roles for old molecules. *Journal of Integrative Plant Biology* 52:98-111.

Calderón-Montaño JM, Burgos-Morón E, Pérez-Guerrero C and López-Lázaro MA (2011) Review on the Dietary Flavonoid Kaempferol. *Mini-Reviews in Medicinal Chemistry* 11:298-344.

Carmona M, Zalacain A and Alonso GL (2006a) The Chemical Composition of Saffron: Colour, Taste and Aroma, 1st ed. Albacete, Bomarzo.

Carmona M, Zalacain A, Sánchez AM, Novella JL and Alonso GL (2006b) Crocetin esters, picrocrocin and its related compounds present in *Crocus sativus* stigmas and *Gardenia jasminoides* fruits. Tentative identification of seven new compounds by LC-ESI-MS. *Journal of Agricultural and Food Chemistry* 54:973-979.

Carmona M, Sánchez AM, Ferreres F, Zalacain A, Tomás-Barberán F and Alonso GL (2007) Identification of the flavonoid fraction in saffron spice by LC/DAD/MS/MS: Comparative study of samples from different geographical origins. *Food Chemistry* 100:445-450.

Comtrade (2015) United Nations Commodify Trade Statistics Database. http://comtrade.un.org/db/dqBasicQueryResults.aspx?cc=091020&px=H0&r=724&y=2014 [20th December 2015].

Del Campo CP, Garde-Cerdán T, Sánchez AM, Maggi L, Carmona M and Alonso GL (2009) Determination of free amino acids and ammonium ion in saffron (*Crocus sativus* L.) from different geographical origins. *Food Chemistry* 114:1542-1548.

Erol O, Can L and Küçüker O (2014) *Crocus yaseminiae* (*Iridaceae)* a new species from South Anatolia, Turkey. *Phytotaxa* 188:103-111.

Erol O, Harpke D and Yildirim H (2015) A new *Crocus* L. (*Iridaceae*) species from SE Turkey, based on morphological and molecular data. *Phytotaxa* 239:223-232.

Eurostat (2011) Flowers and ornamental plants. http://ec.europa.eu/eurostat, data 2011 [August 2015].

Fatehi M, Rashidabady T and Fatehi-Hassanabad Z (2003) Effects of *Crocus sativus* petals' extract on rat blood pressure and on responses induced by electrical field stimulation in the rat isolated vas deferens and guinea-pig ileum. *Journal of Ethnopharmacology* 84:199-203.

Festuccia C, Mancini A, Gravina GL, Scarsella L, Llorens S, Alonso GL, Tatone C, Di Cesare E, Jannini EA, Lenzi A, D' Alessandro AM and Carmona M (2014) Antitumor effects of saffron-derived carotenoids in prostate cancer cell models. *BioMed Research International* 4:1-12.

García-Rodríguez MV, Serrano-Díaz J, Tarantilis PA, López-Córcoles H, Carmona M and Alonso GL (2014) Determination of saffron quality by high-performance liquid chromatography. *Journal of Agricultural and Food Chemistry* 62:8068-8074.

GBIF (2013) GBIF Secretariat: GBIF Backbone Taxonomy, 2013-07-01. http://www.gbif.org/dataset/d7dddbf4-2cf0-4f39-9b2a-bb099caae36c on 2016-04-07.

Goupy P, Vian MA, Chemat F and Caris-Veyrat C (2013) Identification and quantification of flavonoles, anthocyanins and lutein diesters in tepals of *Crocus sativus* by ultra performance liquid chromatography coupled to diode array and ion trap mass spectrometry detections. *Industrial Crops and Products* 44:496-510.

Hadizadeh F, Khalili N, Hosseinzadeh H and Khair-Aldine R (2003) Kaempferol from saffron petals, Iran. *Journal of Pharmaceutical Research* 2:251-252.

Harborne JB and Williams CA (1984) 6-Hydroxyflavones and other flavonoids of *Crocus*. *Zeitschrift für Naturforschung C* 39:18-23.

Harpke D, Meng S, Kerndorff H, Rutten T and Blattner FR (2013) Phylogeny of *Crocus* (*Iridaceae*) based on one chloroplast and two nuclear loci: ancient hybridization and chromosome number evolution. *Molecular Phylogenetics Evolution* 66:617-627.

Harpke D, Peruzzi L, Kerndorff H, Karamplianis T, Constantinidis T, Ranđelović V, Ranđelović N, Jušković M, Pasche E and Blattner FR (2014) Phylogeny, geographic distribution, and new taxonomic circumscription of the *Crocus reticulatus* species group (*Iridaceae*). *Turkish Journal of Botany* 38:1182-1198.

Harpke D, Carta A, Tomović G, Randelović V, Randelović N, Blattner FR and Peruzzi L (2015) Phylogeny, karyotype evolution and taxonomy of *Crocus* series Verni (*Iridaceae*). *Plant Systematics and Evolution* 301:309-325.

Hosseinzadeh H and Younesi HM (2002) Antinociceptive and anti-inflammatory effects of *Crocus sativus* L. stigma and petal extracts in mice. *BMC Pharmacology* 2:1–8.

Hosseinzadeh H and Nassiri-Asl M (2013) Avicenna's (Ibn Sina) the Canon of Medicine and Saffron (*Crocus sativus*): A review. *Phytotherapy Research* 27:475-483.

ITAP (2013) El cultivo del azafrán en Castilla-La Mancha. Boletín 88. Available in http://www.itap.es/media/43952/bolet_n_088._azafr_n.pdf [15th April 2016].

Kanakis CD, Tarantilis PA, Tajmir-Riahi HA and Polissiou MG (2007) Crocetin, dimethylcrocetin, and safranal bind human serum albumin: stability and antioxidative properties. *Journal of Agricultural and Food Chemistry* 55:970-977.

Kerndorff H, Pasche E, Blattner FR and Harpke D (2013a) Fourteen New species of *Crocus* (*Liliiflorae*, *Iridaceae*) from West, South-West and South-Central Turkey. *Stapfia* 99:145-158.

Kerndorff H, Pasche E, Blattner FR and Harpke D (2013b) *Crocus biflorus* Miller (*Liliiflorae*, *Iridaceae*) in Anatolia-Part IV. *Stapfia* 99:159-186.

Kerndorff H, Pasche EH and Harpke D (2014) *Crocus isauricus* Siehe ex Bowles (*Liliiflorae*, *Iridaceae*) and its relatives. *Stapfia* 101:3-18.

Kerndorff H, Pasche EH and Harpke D (2015a) *Crocus lyciotauricus* Kerndorff & Pasche (*Liliiflorae*, *Iridaceae*) and its relatives. *Stapfia* 103:67-80.

Kerndorff H, Pasche EH and Harpke D (2015b) The Genus *Crocus* (*Liliiflorae, Iridaceae*): Lifecycle, Morphology, Phenotypic Characteristics and Taxonomical Relevant Parameters. *Stapfia* 103:27-65.

Kim JH and Fujieda K (1991) Studies of the flower color variation in *Hibiscus syriacus* L. *Journal of Korean Society of Horticulture Science* 32:247-255.

Kim TH, Ku SK, Lee IC and Bae JS (2012) Anti-inflammatory effects of kaempferol-3-*O*-sophoroside in human endothelial cells. *Inflammation Research* 61:217-224.

Kubo I, Marshall GT and Hanke FJ (1988) Rotation locular countercurrent chromatography for natural products isolation. In: Countercurrent Chromatography; Mandava, N. B., Ito, Y., Eds.; Marcel Dekker: New York, pp. 493-507.

Kubo I and Kinst-Hori I (1999) Flavonols from saffron flower: Tyrosinase inhibitory activity and inhibition mechanism. *Journal of Agricultural and Food Chemistry* 47:4121-4125.

Li CY and Wu TS (2002) Constituents of the pollen of *Crocus sativus* L. and their tyrosinase inhibitory activity. *Chemical and Pharmaceutical Bulletin* 50:1305-1309.

Li CY, Lee EJ and Wu TS (2004) Antityrosinase principles and constituents of the petals of *Crocus sativus. Journal of Natural Products* 67:437-440.

Licón C, Carmona M, Llorens S, Berruga MI and Alonso GL (2010) Potential healthy effects of saffron spice (*Crocus sativus* L. stigmas) consumption. In: Husaini, A.M., editor. Saffron. Functional plan science and biotechnology 4. Special issue 2. Japan: © 2010 Global Science Books, Ltd, pp. 64-73.

Limtrakul P, Khantamat O and Pintha K (2005) Inhibition of P-glycoprotein function and expression by kaempferol and quercetin. *Journal of Chemotherathy* 17:86-95.

López-Córcoles H, Brasa-Ramos A, Montero-García F, Romero-Valverde M and Montero-Riquelme F (2015) Phenological Growth stages of saffron Plant (*Crocus sativus* L.) according to the BBCH Scale. *Spanish Journal of Agricultural Research* 13:e09SC01.

Lopresti AL and Drummond PD (2014) Saffron (*Crocus sativus*) for depression: a systematic review of clinical studies and examination of underlying antidepressant mechanisms of action. *Human Psychopharmacology Clinical and Experimental* 29:517-527.

Luo H, Daddysman MK, Rankin GO, Jiang BH and Chen YC (2010) Kaempferol enhances cisplatin's effect on ovarian cancer cells through promoting apoptosis caused by down regulation of cMyc. *Cancer Cell International* 10:16.

MAGRAMA (2015) Anuario de Estadística de 2014. Ministerio de Agricultura, Alimentación y Medio Ambiente. Madrid.

Martina G, Gohb E and Neff AW (2002) Evaluation of the developmental toxicity of crocetin on *Xenopus. Food Chemistry Toxicology* 40:959-964.

Mathew B (1982) The *Crocus.* A revision of the genus *Crocus* (*Iridaceae*). B T Batsford Ltd., London.

Maw G (1886) A monograph of the Genus Crocus. Dulau and Co, London.

Mena Barreto Silva FR, Zanatta L, Silva Frederico MJ, Pizzolatti MG and Machado A (2013) Kaempferol and kaempferitrin: nutraceutical compounds contribute to glucose homeostasis by acting at multiple biological sites. In: Kaempferol. Chemistry, natural occurrences and health benefits, Villers, G. & Fougere, Y., Eds. Nova Science Publishers, Inc., pp. 39-44.

Mo Y, Nagel C and Taylor LP (1992) Biochemical complementation of chalcone synthase mutants defines a role for flavonols in functional pollen. *Proceedings of the National Academy of Sciences of the United States of America* 89:7213-7217.

Moshiri E, Akhondzadeh AB, Noorbala AA, Jamshidi AH, Abbasi SH and Akhondzadeh S (2006) *Crocus sativus* L. (petal) in the treatment of mild-tomoderate depression: A doubleblind, randomized and placebo-controlled trial. *Phytomedicine* 13:607-611.

Negbi M, Dagan B, Dror A and Basker D (1989) Growth, flowering, vegetative reproduction, and dormancy in the saffron (*Crocus sativus* L.). *Israel Journal of Botany* 3:95-113.

Noorbala AA, Akhondzadeh S, Tahmacebi-Pour N and Jamshidi AH (2005) Hydroalcoholic extract of *Crocus sativus* L. versus fluoxetine in the treatment of mild to moderate depression: a double-blind, randomized pilot trial. *Journal of Ethnopharmacology* 97:281-284.

Nørbæk R and Kondo T (1999a) Flavonol glycosides from flowers of *Crocus speciosus* and *C. antalyensis*. *Phytochemistry* 51:1113-1119.

Nørbæk R, Nielsen JK and Kondo T (1999b) Flavonoids from flowers of two *Crocus chrysanthus-biflorus* cultivars: 'Eye-catcher' and 'Spring Pearl' (*Iridaceae*). *Phytochemistry* 51:1139-1146.

Nørbæk R, Brandt K, Nielsen JK, Ørgaard M and Jacobsen N (2002) Flower pigment composition of *Crocus* species and cultivars used for a chemotaxonomic investigation. *Biochemical Systematics and Ecology* 30:763-791.

OJ (1999) Commission regulation (CE) Nº 378/1999 of 19/02/1999. Official Journal of the European Communities of 20/02/1999, L46, p. 13.

OJ (2001) Commission regulation (CE) Nº 464/2001 of 07/03/2001. Official Journal of the European Communities of 08/03/2001, L66, p. 29.

OJ (2005) Commission regulation (CE) Nº 205/2005 of 04/02/2005. Official Journal of the European Union of 05/02/2005, L33, p. 6.

OJ (2009) Commission regulation (CE) Nº 98/2009 of 02/02/2009. Official Journal of the European Union of 03/02/2009, L33, p. 8.

Petersen G, Seberg O, Thorsøe S, Jørgensen T and Mathew B (2008) A phylogeny of the genus *Crocus (Iridaceae)* based on sequence data from five plastid regions. *Taxon* 5:487-499.

Price JR, Robinson GM and Robinson R (1938) Occurrence of kaempferol in *Crocus*. *Journal of the Chemical Society* 281-282.

Roskov Y, Abucay L, Orrell T, Nicolson D, Flann C, Bailly N, Kirk P, Bourgoin T, De Walt RE, Decock W and De Wever A (2016) Species 2000 & ITIS Catalogue of Life, 25th March 2016. Digital resource at www.catalogueoflife.org/col. Species 2000: Naturalis, Leiden, the Netherlands.

Rubio-Moraga A, Mozos AT, Ahrazem O and Gómez-Gómez L (2009) Cloning and characterization of a glucosyltransferase from *Crocus sativus* stigmas involved in flavonoid glucosylation. *BMC Plant Biology* 9:109.

Rubio-Moraga A, Ahrazem O, Rambla JL, Granell A and Gomez-Gomez L (2013) Crocins with high levels of sugar conjugation contribute to the yellow colours of early-spring flowering crocus tepals. *PLOS One* 8:e71946.

Saleem S, Ahmad M, Ahmad AS, Yousuf S, Ansari MA, Khan MB, Ishrat T and Islam F (2006) Effect of saffron (*Crocus sativus*) on neurobehavioral and neurochemical changes in cerebral ischemia in rats. *Journal of Medicinal Food* 9:246-253.

Sánchez AM, Jerz G, Serrano-Díaz J, Alonso GL and Winterhalter P (2011) Flavonol composition and isolation of kaempferol 3-sophoroside from saffron (*Crocus sativus* L.) floral bio-residues. Proceeding of the 40. Deutscher Lebensmittelchemikertag 231; 2011 Sept 12-14; Halle, Germany.

Sánchez AM, Jerz G, Alonso GL and Winterhalter P (2014) Screening and Fractionation of Polyphenols from Saffron (*Crocus sativus* L.) Floral Bio-residues with Preparative HSCCC and Sequential Off-Line Injection to ESI-MS/MS. Proceedings of the 8th International Conference on Countercurrent Chromatography, p. 140.

Sánchez-Vioque R, Rodríguez-Conde MF, Reina-Ureña JV, Escolano-Tercero MA, Herraiz-Peñalver D and Santana-Méridas O (2012) In vitro antioxidant and metal chelating properties of corm, tepal and leaf from saffron (*Crocus sativus* L.). *Industrial Crops and Products* 39:149-153.

Sariri R, Sabbaghzadeh R and Poumohamad F (2011) In-vitro antioxidant and antityrosinase activity of methanol extracts from *Crocus sativus* flowers. *Pharmacologyonline* 3:1-11.

Scheffeldt P and Hrazdina G (1978) Co-pigmentation of anthocyanins under physiological conditions. *Journal of Food Science* 43:517-520.

Serrano-Díaz J, Sánchez AM, Maggi L, Martínez-Tomé M, García-Diz L, Murcia MA and Alonso GL (2012) Increasing the applications of *Crocus sativus* flowers as natural antioxidants. *Journal Food Science* 77:C1162-C1168.

Serrano-Díaz J (2013) Valorization of floral bio-residues from saffron spice production for food applications. Doctoral Thesis. University of Castilla-La Mancha.

Serrano-Díaz J, Sánchez AM, Martínez-Tomé M, Winterhalter P and Alonso GL (2013) A contribution to nutritional studies on *Crocus sativus* flowers and their value as food. *Journal of Food Composition and Analysis* 31:101-108.

Serrano-Díaz J, Sánchez AM, Martínez-Tomé M, Winterhalter P and Alonso GL (2014) Flavonoid determination in the quality control of floral bioresidues from *Crocus sativus* L. *Journal of Agricultural and Food Chemistry* 62:3125-3133.

Smolskaite L, Talou T, Fabre N and Venskutonis P (2011) Valoritation of saffron industry by-products: bioactive compounds from leaves. FOODBALT 2011. Conference proceedings 6 th Baltic Conference on Food Science and Technology "Innovations for Food Science and Productio." 5-6 may, 2011, Jelgava, Latvia. http://llufb.llu.lv/conference/foodbalt/2011/ [28th March 2016].

Spiller GA (1986) Suggestions for a basis on which to determine a desirable intake of dietary fibre. In: Spiller, G.A. (Ed.), CRC Handbook of Dietary Fibre in Human Nutrition. CRC Press, Florida, pp. 281-283.

Stark D (1982) Anatomical and physiological studies of floral tube elongation of *Crocus vernus* (*Iridaceae*). *American Journal of Botany* 69:1476-1482.

Straubinger M, Bau B, Eckestein S, Jezussek M and Winterhalter P (1998) Isolation of new saffron constituents using countercurrent chromatography. In: *Natural Product Analysis*; Schreier, P., Herderich, M., Humpf, H. U., Schwab, W., Eds.; Vieweg: Braunschweig/Wiesbaden, Germany, pp. 27-34.

Straubinger M, Jezussek M, Waibel R and Winterhalter P (1997) Two kaempferol sophorosides from *Crocus sativus*. *Natural Product Letters* 10:213-216.

Takemura T, Takatsu Y, Kasumi M, Marubashi W and Iwashina T (2005) Flavonoids and their distribution patterns in the flowers of *Gladiolus* cultivars. *Acta Horiculturae* 673:487-493.

The Iran Project (2015) http://theiranproject.com/blog/2015/05/11/iran-accounts-for-94-of-world-saffron-production/ [16th December 2015].

Trapero A, Ahrazem O, Rubio-Moraga A, Jimeno ML, Gómez MD and Gómez-Gómez L (2012) Characterization of glucosyltransferase enzyme involved in the formation of kaempferol and quercetin sophorosides in *Crocus sativus*. *Plant Physiology* 159:1335-1354.

Tuberoso CIG, Rosa A, Montoro P, Fenu MA and Pizza C (2016) Antioxidant activity, cytotoxic activity and metabolic profiling of juices obtained from (*Crocus sativus* L.) floral by-products. *Food Chemistry* 199:18-27.

Ulbricht C, Conquer J, Costa D, Hollands W, Iannuzzi C, Isaac R, Jordan JK, Ledesma N, Ostroff C, Grimes Serrano JM, Shaffer MD and Varghese M (2011) An evidence-based systematic review of saffron (*Crocus sativus*) by the natural standard research collaboration. *Journal of Dietary Supplements* 8:58-114.

Ute N, Suzanne PM, Lynne RW, Brian EH and Laurence NO (2007) Flavonols and pancreatic cancer risk. *American Journal of Epidemiology* 166:924-931.

Valizadeh R (1988) Using saffron leaves for animal feeding. Project Report, Scientific and Industrial Research Organization of Iran, Khorasan Center.

Valizadeh R (1989) Evaluation of saffron leaves as animal feed. Iranian Industrial and Scientific Research Organization -Khorasan Center. Technical Report.

Verma SK and Bordia A (1998) Antioxidant property of saffron in man. *Indian Journal of Medical Science* 52:205-207.

Wani BA, Manza AKR and Mohiddin FA (2001) Saffron: a repository of medicinal properties. *Journal of Medicinal Plants Research* 5:2131-2135.

Wiersema JH and León B (2013) World economic plants: a standard reference. CRC press.

Williams CA, Harborne JB and Goldblatt P (1986) Correlations between phenolic patterns and tribal classification in the family *Iridaceae*. *Phytochemistry* 25:2135-2154.

Winkel-Shirley B (2001) Flavonoid biosynthesis. A colorful model for genetics, biochemistry, cell biology, and biotechnology. *Plant Physiology* 126:485-493.

Winkel-Shirley B (2002) Biosynthesis of flavonoids and effects of stress. *Current Opinion in Plant Biology* 5:218-223.

Yüzbaşioğlu S, Aslan S and Özhatay N (2015) *Crocus thracicus* (*Iridaceae*), a new species from north-western Turkey. *Phytotaxa* 197:207-214.

Zeka K, Ruparelia KC, Continenza MA, Stagos D, Vegliò F and Arroo RRJ (2015) Petals of *Crocus sativus* L. as a potential source of the antioxidants crocin and kaempferol. *Fitoterapia* 107:128-134.

Zhang Z, Wang CZ, Wen XD, Shoyama Y and Yuan CS (2013) Role of saffron and its constituents on cancer chemoprevention. *Pharmaceutical Biology* 51:7.

Zheng CJ, Li L, Ma WH, Han T and Qin LP (2011) Chemical constituents and bioactivities of the liposoluble fraction from different medicinal parts of *Crocus sativus*. *Pharmaceutical Biology* 49:756-763.

In: Kaempferol ISBN: 978-1-63485-828-1
Editors: T.Garde-Cerdán and A.G. Diago ©2016 Nova Science Publishers, Inc.

Chapter 6

SAFFRON CROCUS (*CROCUS SATIVUS* L.) AS A SOURCE OF KAEMPFEROL

Keti Zeka[1,*] and Randolph Arroo[2]
[1]University of L'Aquila, Department of Life, Health and Environmental Sciences, Coppito (AQ), Italy
[2]De Montfort University, Leicester School of Pharmacy, The Gateway, Leicester, UK

ABSTRACT

Flavonols have been identified in a wide range of fruits and vegetables, and a flavonol-rich diet has been linked with a relatively low occurrence of degenerative diseases like rheumatoid arthritis, Alzheimer's disease, and various forms of cancer. This knowledge has resulted in the development of various food supplements and nutraceutical products, and given rise to a very profitable branch of industry. Notably quercetin preparations have become popular, mainly because of the wide availability and modest cost of this flavonol. Kaempferol has better pharmacokinetic properties, but is roughly 1000 times more expensive than the popular food supplement quercetin. Kaempferol and its glycosides are gaining increasing interest as for their antioxidant activity as a food supplement, in functional foods, beverages drinks, in pharmaceutical preparations and cosmetic formulations. Recently, it was demonstrated that kaempferol can be extracted in good

* E-mail: keti.zeka@graduate.univaq.it.

quantities from the petals of the saffron crocus (*Crocus sativa* L.). Thus, rather than a waste product of the saffron spice production, the petals can be a readily exploitable good source of kaempferol for many applications, ranging from nutraceutical food supplements to cosmetic anti-ageing creams.

1. Introduction

The use of natural products is increasingly becoming a popular strategy to treat or prevent disease within industrialized, Western nations (Coulter and Willis, 2004). The products are used as an alternative medicine and taken as crude herbal preparations or tinctures. The general perception is that natural products present no risk and may help (Seely and Oneschuk, 2008). The use of herbal remedies has been documented in many different patient groups as well as the general population to promote healthly benefits (Carpenter et al., 2008; Hess et al., 2008; Lapi et al., 2008; Zhang et al., 2008). An increasing number of health professionals recommend over-the-counter (OTC) and natural health products to their patients. The desire for more 'natural' alternatives coupled with the requirement to inform patients of all possible side effects and warnings for conventional pharmaceuticals may partially explain the reported increase in the use of herbal remedies (Holst et al., 2011).

On a global scale, the high costs of western synthetic pharmaceuticals put modern health care services out of reach of most of the world's population. Thus, this part of the population must rely on traditional medicine and medicinal plants to meet their primary health care needs. The World Health Organization (WHO) has estimated that about 80% of the world's inhabitants rely on traditional medicines for their primary health care needs, and most of these therapies involve the use of plant extracts or their active components. Even where modern medical care is available and affordable, many people prefer more traditional practices. This is particularly true for First Nations and immigrant populations, who have tended to retain ethnic medical practices.

The investigation of natural products as a source of novel human therapeutics reached its peak in the Western pharmaceutical industry during 1970-1980, and resulted in a pharmaceutical landscape heavily influenced by non-synthetic molecules (Vazhayil et al., 2014; Patwardhan and Marhelkar, 2009). However, "natural" does not necessarily means "safe." Some herbal products are extremely effective but so dangerous that they should only be used in the hands of skilled medical professionals. Some of the most effective medicines that were developed in this era include derivatives of highly toxic natural products that are now in clinical use in cancer chemotherapy, e.g., paclitaxel, vinca alkaloids, podophyllotoxin derivatives, camptothecin and its derivatives (Bhanot et al., 2011).

More recently, fruits, vegetables, and spices have attracted considerable attention from both the scientific community and the general public for their potential ability to suppress cancer diseases. A diet rich in fruits, vegetables and certain spices is correlated with a lower incidence of cancers (WCRF/IACR, 2007). Arguably, health-enhancing compounds found in food plants can be generally recognized as safe and may be exempted from the usual Federal Food, Drug, and Cosmetic Act (FFDCA) food additive tolerance requirements. In the wake of this interest in food-derived natural products, a number of companies have now invested in the search for plant-based drugs. In fact, improvements in automation and robotics have facilitated laboratory evaluation of large amounts of samples in a short time. Particularly, plant products with antioxidant properties and compounds with mild oestrogenic activity – so-called phytoestrogens - have attracted the interest of complementary health and cosmetic industries. These products are linked with anti-inflammatory activity (Bobe et al., 2010) and thus with prevention of a variety of degenerative diseases, and they can be marketed as 'natural' and safe. In the field of cosmetics, they are perceived to be less aggressive for the skin than man-made chemicals.

2. Medicinal Uses of Saffron

Among the health-enhancing natural products, saffron spice has always played a special role. Saffron is the dried stigma of the saffron crocus, *Crocus sativus* L. (Iridaceae), and is the most expensive and famous spice that exists. In autumn, the plant flowers, and the flowers are collected early in the morning when they are just opening. The manual separation of three stigmas from each flower is one of the most critical point of saffron processing. The time lost in this operation and the handling time required – which depends on the farmer's experience - may be a key determining factor affecting saffron quality and levels of contamination. Though alternatives to this time-consuming process have been proposed - e.g., performing the manual separation using a vertical air column (Emadi, 2009) – much of this work is still done traditionally in cottage industry.

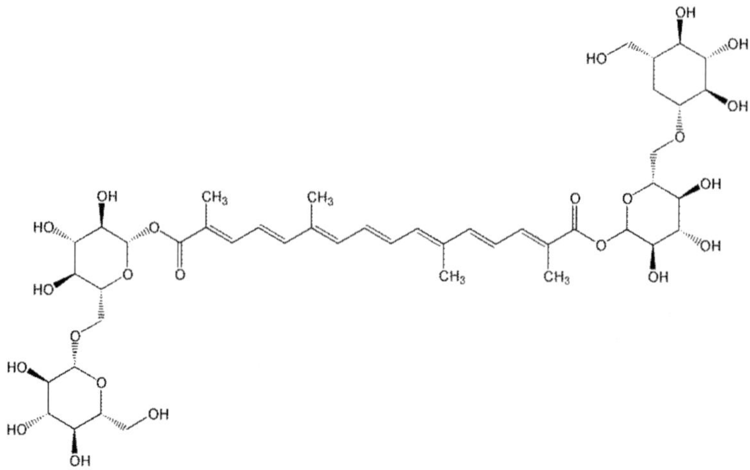

Figure 1. Structure of α-crocin.

Saffon's importance has been known since ancient times. In traditional medicine is used to treat many diseases (Sharafzadeh, 2011). In medicine, it is used as an antispasmodic, eupeptic, gingival sedative, anticatarrhal, nerve sedative, diaphoretic, expectorant, stimulant, stomachic, aphrodisiac and emmenagogue (Rios et al., 1996). Considering all medical-biological activities reported for saffron spice in many papers, it is possible to make a grouping (Abdullaev and Espinosa-Aguirre, 2004) effect on learning behaviour, effect on blood pressure, anticonvulsant effect, antinociceptive and anti-inflammatory effects, mutagenic or antimutagenic effects, antigenotoxic effect, tumoricidal effect, and cytotoxic effect.

Saffron and saffron extracts have a very low toxicity *in vivo* studies; adverse effects for humans have been reported for 5 g and above (Schmidt et al., 2007). This is equivalent to approximately 1,500 stigmas. Older literature sources have mentioned nausea and revulsion at doses between 1.3 g and 2 g, but these are likely to refer to another plant, the 'meadow saffron' *Colchium autumnale* L. (Colchicaceae), a plant species that – in spite of present similar flowers, that also show up in autumn - is not related to *Crocus sativus* (Schmidt et al., 2007).

Several papers mention crocins as the most abundant compounds in saffron stigma. Crocins present deep red colour, and are dissolved in water form an orange coloured solution, making them widely used as natural food colorant. Crocins constitute approximately 6 to 16% of the total dry matter of saffron stigmas, depending upon the variety, growing conditions, and processing methods (Gregory, Menary, and Davies, 2005). Crocin 1 (or α-crocin), a digentiobioside carotenoid, is the most abundant crocin with a high solubility due to its sugar moieties.

Most of the saffron literature is based on the study of its stigmas, possibly because these are highly valued as spice. Only a small part examines *Crocus sativus* petals. The main reason for this may be that petals have never been considered to be of any particular value. In fact, saffron petals are considered a waste product, being discarded or just used as compost fertilizer. However, they are known for containing flavonoids and anthocyanins. Petals, entire flowers, and stamens of *C. sativus*, represent an interesting source of phenolic compounds. The high variety of different flavonoids, included kaempferol and the large amounts of crocins confirm the importance of the whole saffron flower (Montoro et al., 2012).

An LC-DAD-MS (ESI+) analysis of *Crocus sativus* petals, showed that the fractions that were the richest in flavonoids and flavonoid glycosides, demonstrated the strongest antioxidant capacity. The antioxidant capacity of the petals is higher than that of the stigmas (Termentzi and Kokkalou, 2008).

Applying three model systems - i.e., β-carotene/linoleic acid, DPPH, and reducing power through potassium ferricyanide - the antioxidant activity was found to be correlated with the level of phenolics in a methanolic extract of saffron petals (Goli et al., 2012).

A hydro alcoholic extract of *C. sativus* petals at 20 mg/kg improves the effect of acetaminophen-induced hepatotoxicity in male Wistar rats, and returned blood parameters and the histopathology of liver to almost normal levels. This effect is associated with the high flavonoid content and the antioxidant properties of the extract (Omidi et al., 2014).

A study about the effects of saffron petal extract on blood parameters, immune system, and spleen histology, using five treatments that were used in a completely randomized design, indicated that a dose of 75 mg/kg causes an increase in antibody response without any change in haematological parameters and spleen histology (Babaei et al., 2014).

Saffron petals are a promising source of antioxidants. Considering that oxidative processes and the formation of oxygen radicals are the main cause of macromolecules biological deterioration, which plays a role in the process of ageing and development of degenerative disorders (Sanchez-Moreno et al., 1999; Verrastro et al., 2015; Wojtunik-Kuleza et al., 2016); the demonstration of antioxidant activity in saffron petals may be considered important. Thus, chemicals and nutritional compounds present in the petals of saffron may be a source of new dermal treatments both medical and cosmetic.

3. FLAVONOLS

Flavonols are among the most abundant antioxidants in food plants. Flavonols are a subgroup of a larger group of structurally related compounds, the flavonoids, whose basic structure consists of two phenyl groups joined by a three carbon bridge (Bohm 1998; Beecher 2003). Flavonoids are commonly found in plants and constitute a significant part of the human diet (Wojdylo et al., 2007).

The biosynthesis of flavonoids occurs by condensation of 4-coumaroyl-CoA (C6-C3) with three molecules of malonyl-CoA (C6) (Winkel-Shirley, B. 2001, 2002). At this point, two enzymes: chalcone synthase (EC 2.3.1.74) and chalcone isomerase (EC 5.5.1.6) catalyse first the formation of narigenin chalcone (C6-C3-C6), and secondly the flavanone naringenin.

The enzyme flavanone 3- dioxygenase (EC 1.14.11.9) introduces a hydroxyl group in naringenin at C3 to form dihydrokaempferol, and in last step, at C2-C3 of the dihydrokaempferol are introduced double bond from the enzyme flavonol synthase (EC 1.14.11.23) to produce kaempferol (Calderón-Montaño et al., 2011).

This biosynthetic scheme is common to most of the plant families, and many further biosynthetic pathways following this generic scheme provide a large variety of structurally related compounds. Within the flavonols, new compounds can be formed by modifying the aromatic rings, e.g., with the introduction of hydroxyl groups, or methylation of hydroxyl groups to form methyl ethers. Another main way to create more diverse molecules is by linking one, or more sugars to the hydroxyl groups of the core flavonol, resulting in the formation of glycosides. The variety of glycosides is staggering, only for the flavonol kaempferol hundreds glycosidic forms have been identified in the plant kingdom (Harborne and Williams, 1988; Williams and Grayer, 2004). A wide variety of plant species accumulate different flavonols, and flavonol glycosides.

The major dietary sources of flavonols vary according to the geographical area and culture. Red wine is the predominant dietary source of flavonoles in Italy, France or Spain and tea is the major one in the Japanese and British cultures. In U.S.A., Finland, Greece, and former Yugoslavia, onions and apples are among the foods consumed with higher content of flavonols (de Vries et al., 1998).

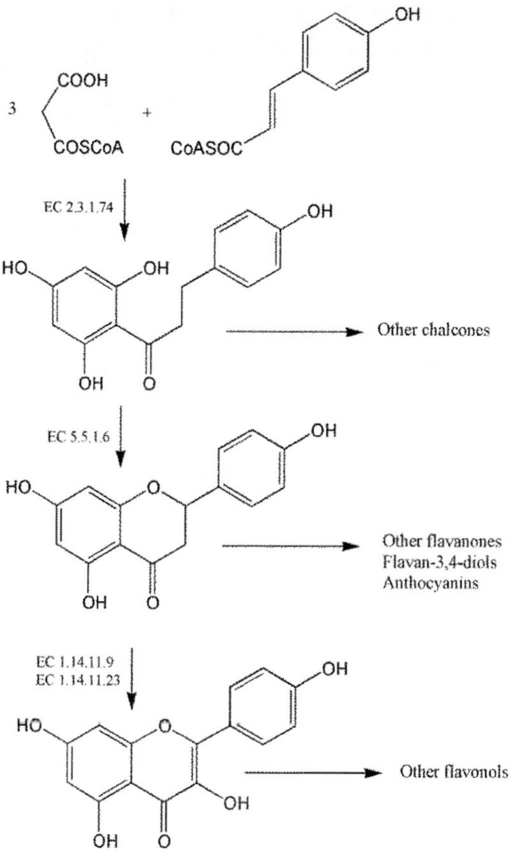

Figure 2. Flavonol biosynthesis.

Kaempferol is found in apples, grapes, tomatoes, green tea (Kim and Choi, 2013), potatoes (Liu, RH. 2013), onions, broccoli, Brussels sprouts, squash, cucumbers, lettuce, green beans, peaches, blackberries, raspberries, spinach (Calderón-Montaño et al., 2011), and strawberries (Somerset and Johannot, 2008). Other plants that are known to contain kaempferol include *Aloe vera*, *Coccinia grandis*, *Euphorbia pekinensis*, *Glycine max*, *Hypericum perforatum*, *Rosmarinus officinalis*, *Sambucus nigra* (Calderón-Montaño et al., 2011), *Cuscuta chinensis* (Donnapee et al., 2014), *Moringa oleifera* (Anwar et al., 2007). Saffron petals are one of the richest sources of kaempferol and its glycosides. The level of this flavonol in saffron petals, in g/kg, is about 100 times higher than that in foods considered 'rich' in kaempferol (Zeka et al., 2015).

The biosynthesis of plant secondary compounds is often tightly regulated. Plants respond to changes in their environment, e.g., the biosynthesis of flavonoids is regulated by ultraviolet light. When dill cell cultures were subjected to UV-B radiation, quercetin-3-*O*-β-glucuronide was synthesized (Mohle et al., 1985). The flavonoids are thought to play a role in protection against damage caused by ultraviolet light (Mariani et al., 2008). We may speculate that the high levels of kaempferol accumulating in saffron petals may be a response to their habitat, i.e., dry mountainous areas with high levels of UV rays from sun light.

4. Purification, Isolation and Identification

Research on saffron has resulted in several patents on extraction techniques, specifically for the extraction of compounds from the valuable stigmas, e.g., hydro-alcoholic extraction to obtain ready-to-use saffron pigments (crocins) and flavour concentrate used as food colorant or in pharmaceutical applications (Agarwal et al., 2006; Garcia Fernandez, 2002). Others studies have focused on the application of safranal and related compounds that capture spicy/saffron-like odour notes for its use as perfumed ingredients (Fehr and Blanc, 2012). A specific application involves partial hydrolysis of α-crocin and other complex crocetin glycosides from saffron stigmas to form crocetin monoesters which are more readily absorbed in the human gut wall, being thus more effective as active pharmaceutical ingredients (Eidenberger, 2010).

When considering extraction of potentially valuable compounds from *C. sativus* petals, similar rules are applied to the extraction from stigmas. The cells of living plants contain not only low molecular-weight compounds but also enzymes that may alter or degrade them. Within the cell, they are compartmentalized in different organelles (e.g., cytoplasm, plastids, vacuole) separated by internal membranes that keep the different constituents apart. When the plant dies, the lipidic bilayer membranes break down and enzymes can start to catalyse various chemical reactions in other cell compartments, e.g., oxidation or hydrolysis. Preservation aims to limit these processes as far as possible. The most common method for preserving plant material is drying since enzyme processes typically take place in aqueous solutions.

Another disadvantage of the presence of moisture is that it would allow microbial, notably fungal, growth, which would lead to a rapid deterioration of plant compounds. Therefore, removal of water from the cell will largely prevent degradation of the cell constituents.

Anthocyanins, which give blueish/purple colour to the saffron petals may be used as food-grade natural colourants. This group of compounds is particularly prone to degradation, and the drying process is critical to their stability. It was found that the traditional drying method – in the sun at ambient temperature - preserves the anthocyanins better than rapid drying at high temperatures in an oven or by means of microwave (Heydari et al., 2014). Furthermore, a gentle extraction of powdered plant material in an aqueous enzymatic mix containing cellulase, hemicellulase, and pectinase yielded higher amounts of anthocyanins than the extraction in ethanol where browning and polymerization occurred (Lotfi et al., 2015). Finally, saffron petal extracted anthocyanins can be freeze dried and stabilized by microencapsulation in maltodextrin or gum Arabic (Khazaei et al., 2014).

Typical extraction methods for flavonols and flavonol glycosides include extraction in a mix of alcohol and water, followed by filtration and concentration of the filtrate under vacuum using a rotary evaporator. Subsequent steps may involve acid hydrolysis to break glycosidic bonds, followed by separation of sugars and the relatively lipophilic flavonol aglycones in a separation funnel, and further purification of the aglycone fraction by flash column chromatography.

Levels of flavonols can be quantified by HPLC, using commonly available C_{18} reverse phase columns, and UV-VIS or diode array detection to measure absorbance at 254 nm. Only few flavonoid glycosides are commercially available for reference purposes, and their direct quantitative analysis is often impractical. In addition, glycosides have short retention times and their peaks on a typical HPLC chromatogram tend to cluster together, but they can be analyzed using LC-MS equipment (Benayad et al., 2014) which is, however, less commonly available to researchers. Considering that flavonol aglycones rather than their glycosides are considered as the active pharmaceutical ingredients from plants, many authors only report the 'total flavonol' content, i.e., the amount of flavonol aglycones after hydrolysis of the glycosides.

An interesting extraction is that of quercetin from onion waste, which is based on extracting quercetin glycosides with water in a pressure cooker at 120°C and 5.0 MPa, followed by enzymatic hydrolysis with water at 90°C using a thermostable β-glucosidase from *Thermotoga neapolitana* (Herrero et al., 2015). Another, more conventional hydrolysis technique involved refluxing plant material in 1 M HCl for over an hour to obtain kaempferol aglycone from saffron petals or aerial parts of *Aconitum anthora* (Zeka et al., 2015; Mariani, 2008), or quercetin, kaempferol, and myricetin from various vegetables (Andarwulan et al., 2010). The flavonol aglycones are relatively stable and not prone to degradation at high temperatures or low pH.

5. Mechanism of Action

'We are what we eat' is a well-known principle in every diet. Diet is recognized as one of the most important factors that influence human metabolism and a diet rich in fruits and vegetables is globally accepted as the most beneficial. Several mechanisms have been proposed that may explain the health enhancing benefits of fruits and vegetables, the most common models are through anti-oxidant activity and through activity of phytoestrogens.

Our metabolism depends on oxidation and, consequently, generates free radicals. Additional free radicals are produced from air pollutants, pesticides, radiation and smoking. High levels of radical oxygen species are referred to as oxidative stress which has been associated with damage of macromolecules in living cells, aging, and degenerative diseases (Aviram, 2000; Ito et al., 2004; Khan et al., 2010). In the human body, physiological mechanisms are present to prevent oxidative stress, e.g., glutathions and superoxide dismutase (EC 1.15.1.). Flavonoids derived from the diet can act as additional antioxidants through the donation of electrons from their double bonds and hydroxyl groups that can stabilize free radicals (Machlin and Bendich, 1987).

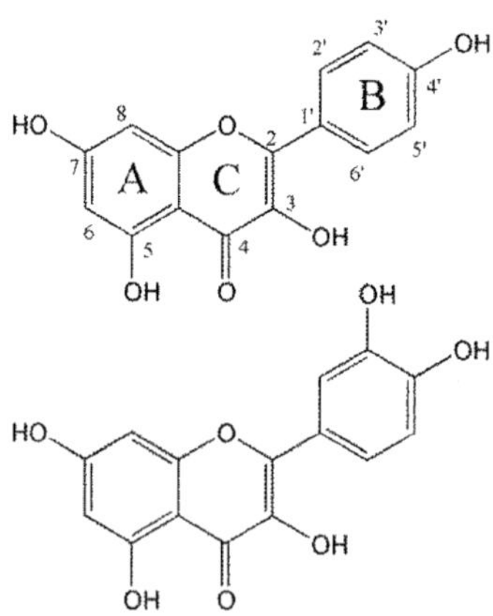

Figure 3. Structures of kaempferol (top) and quercetin (bottom).

Based on the notion that flavonols derived from food products may prevent the onset of degenerative disorders, the complementary and alternative health industry has promoted regular consumption flavonol supplements. Particularly quercetin has been recommended as a supplement with antioxidant properties that may help manage a number of inflammatory health problems, including heart disease and blood vessel problems, allergies, infections, chronic fatigue, and symptoms related to autoimmune disorders like arthritis. However, the scientific literature is rather critical with the consumption of high doses of quercetin. Depending on the doses, flavonols with a catechol moiety in the B-ring, e.g., quercetin or myricetin, can also act as pro-oxidants and thus exacerbate oxidative damage (Kessler et al., 2003; Laughton et al., 1989; Sahu et al., 2001). In addition, flavonols with a catechol motif in their B-ring are prone to further oxidation and can form DNA adducts (Kessler et al., 2003; van der Woude et al., 2006; Walle et al., 2003), thus severely disrupting normal cell processes. Arguably, the amounts quercetin that we obtain through our daily diet do no harm, but excessive intake of quercetin through food supplements may not be without risk.

Kaempferol has not been shown to have pro-oxidative properties or be involved in adduct formation. In addition, this flavonol was found to be stable at a wide range of different pH values, which is in contrast to quercetin where an acid pH is required to prevent its oxidative degradation (Day, 2001; DuPont et al., 2004).

Flavonoids are mostly found in the nature as glycosides. The attached sugar groups, which make their structure more complicated, are the reason why flavonoids are poorly absorbed through the gut wall. Flavonoid glycosides pass through the human digestive system, and are only degraded in the colon by microorganisms able to hydrolyze the glycosidic bond, thus forming flavonoid aglycones that are absorbed at the distal end of the digestive tract (Barve et al., 2009; DuPont et al., 2004; Murota and Terao, 2003).

Only few studies have compared the bioavailability of different flavonols, but their results suggest that kaempferol is absorbed more efficiently than quercetin, and/or that quercetin is more extensively metabolized to other compounds (de Vries et al., 1998; DuPont et al., 2004; Nielsen et al., 1997). Kaempferol is not converted into quercetin by common cytochromes P450 enzymes that control phase I detoxification (DuPont et al., 2004). However, the conversion of kaempferol into quercetin is catalysed by CYP1 enzymes, a group of cytochromes P450 that are known to be selectively expressed in tumours and pre-cancerous cells (Arroo et al., 2009, 2014; Breinholt et al., 2002). The latter observation opens the possibility that levels of kaempferol may rise to relatively high levels without much effect on normal cells, however only in pre-cancerous cells it can be converted into quercetin and disrupt cell physiology. Intriguingly, kaempferol appears to have pro-oxidant properties and induces apoptosis in the promyelocitic human leukemia cell line U937 (Marfe et al., 2009) and glioblastoma cell line U87MG (Sharma et al., 2007) which are both known to express CYP1 enzymes after induction (Matsunawa et al., 2012; Stiborova et al., 2014).

CONCLUSION

Petals of the saffron crocus (*Crocus sativus* L.) are considered a waste product in the production of the valuable spice saffron. Recent analyses have shown that the petals can be a very good source of kaempferol glycosides, revealing them as a promising source of inflammatory agents that may assist in the prevention of degenerative disorders. Several pre-clinical trials have indicated that the therapeutic index of kaempferol is considerably higher than that of the widely used food supplement quercetin. Kaempferol and its glycosides are gaining increasing interest because of their antioxidant activity as a food supplement, in functional foods, beverages, in pharmaceutical preparations and in cosmetic formulations.

ACKNOWLEDGMENTS

The authors would like to thank Professor Maria Adelaide Continenza and Dr Ketan Ruparelia for their support.

REFERENCES

Abdullaev FI and Espinosa-Aguirre JJ (2004) Biomedical properties of saffron and its potential use in cancer therapy and chemoprevention trials. *Cancer Detection and Prevention* 28:426-432.

Agarwal SG, Thappa RK, Agnihotri VK, Suri OP and Qazi GN (2006) Method for the extraction of saffron pigments and flavor concentrate. Patent US-7,070,823.

Andarwulan N, Batari R, Sandrasari DA, Bolling B and Wijaya H (2010) Flavonoid content and antioxidant activity of vegetables from Indonesia. *Food Chemistry* 121:1231–1235

Anwar F, Latif S, Ashraf M and Gilani AH (2007). *Moringa oleifera*: a food plant with multiple medicinal uses. *Phytotherapy Research* 21:17–25.

Arroo RRJ, Androutsopoulos V, Beresford K, Ruparelia K, Surichan S, Wilsher N and Potter GA (2009) Phytoestrogens as natural prodrugs in cancer prevention: Dietary flavonoids. *Phytochemistry Reviews* 8:375-386.

Arroo RRJ, Beresford K, Bhambra AS, Boarder M, Budriesi R, Cheng Z, Micucci M, Ruparelia KC, Surichan S and Androutsopoulos VP (2014) Phytoestrogens as natural prodrugs in cancer prevention: towards a mechanistic model. *Phytochemistry Reviews* 13:853-866.

Aviram M (2000) Review of human studies on oxidative damage and antioxidant protection related to cardiovascular diseases. *Free Radical Research* 33:S85–S97.

Babaei A, Arshami J, Haghparast A and Mesgaran MD (2014) Effects of saffron (*Crocus sativus*) petal ethanolic extract on hematology, antibody response, and spleen histology in rats. *Avicenna Journal Phytomedicine* 4:103–109.

Benayad Z, Gómez-Cordovés C and Es-Safi NE (2014) Characterization of Flavonoid Glycosides from Fenugreek (*Trigonella foenum-graecum*) Crude Seeds by HPLC–DAD–ESI/MS Analysis. *International Journal of Molecular Sciences* 15:20668-20685.

Barve A, Chen C, Hebbar V, Desiderio J, Saw CL and Kong AN (2009) Metabolism, oral bioavailability and pharmacokinetics of chemopreventive kaempferol in rats. *Biopharmaceutics & Drug Disposition* 30:356–365.

Bhanot A, Sharma R and Noolvi MN (2011) Natural sources as potentian anti-cancer agents: A review. *International Journal of Phytomedicine* 3:9-26.

Beecher GR (2003) Overview of dietary flavonoids: nomenclature, occurrence and intake. *The Journal of Nutrition* 133:3248S-3254S.

Bobe G, Albert PS, Sansbury LB, Lanza E, Schatzkin A and Colburn NH (2010). Interleukin-6 as a potential indicator for prevention of high risk adenoma recurrence by dietary flavonols in the polyp prevention trial. *Cancer Prevention Research (Phila)* 3:764–775.

Bohm BA (1998) Introduction to flavonoids. Amsterdam: Harwood Academic Publishers.

Breinholt VM, Offord EA, Brouwer C, Nielsen SE, Brosen K and Friedberg T (2002) In vitro investigation of cytochrome P450-mediated metabolism of dietary flavonoids. *Food and Chemical Toxicology* 40:609–616.

Calderón-Montaño JM, Burgos-Morón E, Pérez-Guerrero C and López-Lázaro M (2011) A Review on the Dietary Flavonoid Kaempferol. *Mini-Reviews in Medicinal Chemistry* 11:298-344.

Carpenter CL, Ganz PA and Bernstein L (2008) Complementary and alternative therapies among very long-term breast cancer survivors. *Breast Cancer Research and Treatment* 116:387–396.

Coulter ID and Willis EM (2004) The rise and rise of complementary and alternative medicine: a sociological perspective. *Medical Journal of Australia* 180:587–589.

Day AJ (2000) Human absorption and metabolism of flavonoid glycosides. PhD dissertation, University of East Anglia, UK.

DeVries JHM, Hollman PCH, Meyboom S, Buysman MNCP, Zock PL, van Staveren WA and Katan MB (1998) Plasma concentrations and urinary excretion of the antioxidant flavonols quercetin and kaempferol as biomarkers for dietary intake. *American Journal of Clinical Nutrition* 68:60–65.

Donnapee S, Li J, Yang X, Ge AH, Donkor PO, Gao XM and Chang YX (2014) Cuscuta chinensis lam.: A systematic review on Ethnopharmacology, phytochemistry and pharmacology of an important traditional herbal medicine. *Journal of Ethnopharmacology* 157:292–308.

DuPont MS, Day AJ, Bennett RN, Mellon FA and Kroon PA (2004) Absorption of kaempferol from endive, a source of kaempferol-3-glucuronide, in humans. *European Journal of Clinical Nutrition* 58:947–954.

Eidenberger T (2010) Hydrolysate of crocin. Patent US 8569247 B2.

Emadi B (2009) Separating saffron flower parts using vertical air column. *International Journal of Biological and Life Sciences* 5:1.

Fehr C and Blanc PA (2012) Perfuming ingredients with saffron odor. Patent US 8,222,199 B2.

Fernandez Garcia E (2002) Process for obtaining an extract of saffron. Patent US 6,458,399 B1.

Goli SAH, Mokhtari F and Rahimmalek M (2012) Phenolic Compounds and Antioxidant Activity from Saffron (*Crocus sativus* L.) Petal. *Journal of Agricultural Science* 4:10.

Gregory MJ, Menary RC and Davies NW (2005) Effect of drying temperature and air flow on the production and retention of secondary metabolites in saffron. *Journal of Agricultural and Food Chemistry* 53:5969–5975.

Harborne JB and Williams CA (1988) Flavone and flavonol glycosides. In: The Flavonoids: Advances in Research since 1980 (J. B. Harborne ed.) Chapman & Hall.

Herrero M, del Pilar Sánchez-Camargo A, Cifuentes A and Ibáñez E (2015) Plants, seaweeds, microalgae and food by-products as natural sources of functional ingredients obtained using pressurized liquid extraction and supercritical fluid extraction. *Trends in Analytical Chemistry* 71:26–38.

Hess S, De Geest S, Halter K, Dickenmann M and Denhaerynck K (2008) Prevalence and correlates of selected alternative and complementary medicine in adult renal transplant patients. *Clinical Transplantation* 23:56–62.

Heydari S, Rezaei R and Haghayegh GH (2014) Effect of drying processes on stability of anthocyanin extracts from saffron petal. *Evolving Trends in Engineering and Technology* 2:13-18.

Holst L, Wright D, Haavik S and Nordeng H (2011) Safety and efficacy of herbal remedies in obstetrics-review and clinical implications. *Midwifery* 27:80–86.

Ito K, Nakazato T, Yamato K, Miyakawa Y, Yamada T and Hozumi N (2004) Induction of apoptosis in leukemic cells by homovanillic acid derivative, capsaicin, through oxidative stress: Implication of phosphorylation of p53 at Ser-15 residue by reactive oxygen species. *Cancer Research* 64:1071–1078.

Kessler M, Ubeaud G and Jung L (2003) Anti- and pro-oxidant activity of rutin and quercetin derivatives. *Journal of Pharmacy and Pharmacology* 55:131–142.

Khan MA, Tania M, Zhang D and Chen H (2010) Antioxidant enzymes and cancer. *Chinese Journal of Cancer Research* 22:87–92.

Khazaei KM, Jafaria SM, Ghorbani M and Kakhki AH (2014) Application of maltodextrin and gum Arabic in microencapsulation of saffron petal's anthocyanins and evaluating their storage stability and color. *Carbohydrate Polymers* 105:57–62.

Kim SH and Choi KC (2013) Anti-cancer Effect and Underlying Mechanism(s) of Kaempferol, a Phytoestrogen, on the Regulation of Apoptosis in Diverse Cancer Cell Models. *Toxicological Research* 29:229–234.

Lapi F, Vannacci A and Moschini M (2008) Use, attitudes and knowledge of complementary and alternative drugs (CADs) among pregnant women: a preliminary survey in Tuscany. *Evidence-Based Complementary & Alternative Medicine* 7:477–486.

Laughton MJ, Halliwell B, Evans PJ and Hoult JRS (1989) Antioxidant and pro-oxidant actions of the plant phenolics quercetin, gossypol and myricetin - effects on lipid-peroxidation, hydroxyl radical generation and bleomycin-dependent damage to DNA. *Biochemical Pharmacology* 38:2859-2865.

Liu RH (2013) Health-promoting components of fruits and vegetables in the diet. *Advances in Nutrition* 4:384S–392S.

Lotfi L, Kalbasi-Ashtari A, Hamedi M and Ghorbani F (2015) Effects of enzymatic extraction on anthocyanins yield of saffron tepals (Crocus sativus) along with its color properties and structural stability. *Journal of Food and Drug Analysis* 23:210–218.

Machlin LJ and Bendich A (1987) Free radical tissue damage: protective role of antioxidant nutrients. *FASEB Journal* 1:441-445.

Marfe G, Tafani M, Indelicato M, Sinibaldi-Salimei P, Reali V, Pucci B, Fini M and Russo MA (2009) Kaempferol Induces Apoptosis in Two Different Cell Lines Via Akt Inactivation, Bax and SIRT3 Activation, and Mitochondrial Dysfunction. *Journal of Cellular Biochemistry* 106:643–650.

Mariani C, Braca A, Vitalini S, De Tommasi N, Visioli F and Fico G (2008) Flavonoid characterization and in vitro antioxidant activity of *Aconitum anthora* L. (Ranunculaceae). *Phytochemistry* 69:1220–1226.

Marfe G, Tafani M, Indelicato M, Sinibaldi-Salimei P, Reali V, Pucci B, Fini M and Russo MA (2009) Kaempferol Induces Apoptosis in Two Different Cell Lines Via Akt Inactivation, Bax and SIRT3 Activation, and Mitochondrial Dysfunction. *Journal of Cellular Biochemistry* 106:643–650.

Matsunawa M, Akagi D, Uno S, Endo-Umeda K, Yamada S, Ikeda K and Makishima M (2012) Vitamin D Receptor Activation Enhances Benzo[a]pyrene Metabolism via CYP1A1 Expression in Macrophages. *Drug Metabolism and Disposition* 40:2059-2066.

Möhle B, Heller W and Wellmann E (1985) UV-Induced biosynthesis of quercetin 3-O-β-D-glucuronide in dill cell cultures. *Phytochemistry* 24:465–467.

Montoro P, Maldini M, Luciani L, Tuberoso CIG, Congiu F and Pizza C (2012) Radical Scavenging Activity and LC-MS Metabolic Profiling of

Petals, Stamens, and Flowers of *Crocus sativus* L. *Journal of Food Science* 77:C893–C900.

Murota K and Terao J (2003) Antioxidative flavonoid quercetin: implication of its intestinal absorption and metabolism. *Archives of Biochemistry and Biophysics* 417:12–17.

Nielsen SE, Kall M, Justesen U, Schou A and Dragsted LO (1997) Human absorption and excretion of flavonoids after broccoli consumption. *Cancer Letters* 114:173–174.

Omidi A, Riahinia N, Montazer Torbati MB and Behdani M-A (2014) Hepatoprotective effect of *Crocus sativus* (saffron) petals extract against acetaminophen toxicity in male Wistar rats. *Avicenna Journal of Phytomedicine* 4:330-336.

Patwardhan B and Marhelkar RA (2009) Traditional medicine inspired approaches to drug discovery can Ayurveda show the way forward. *Drug Discovery Today* 14:804-811.

Rios JL, Recio MC, Giner RM and Mhnez S (1996) An Update Review of Saffron and its Active Constituents ." *Phytotherapy Research* 10:189-193.

Sahu SC, Flynn TJ, Bradlaw JA, Roth WL, Barton CN and Yates JG (2001) Pro-oxidant effects of the flavonoid myricetin on rat hepatocytes in culture. *Toxicology Methods* 11:277-283.

Sánchez-Moreno C, Larrauri J and Saura-Calixto F (1999) Free radical scavenging capacity and inhibition of lipid oxidation of wines, grape juices and related polyphenolic constituents. *Food Research International* 32:407-412.

Schmidt M, Betti G and Hensel A (2007) Saffron in phytotherapy: pharmacology and clinical uses. Wiener Medizinische Wochenschrift 157:315-319.

Seely D and Oneschuk D (2008) Interactions of natural health products with biomedical cancer treatments. *Current Oncology* 15:S81-S86.

Sharafzadeh S (2011) Saffron: A Concise Review of Researches. *Advances in Environmental Biology* 5:1617-1621.

Sharma V, Joseph C, Ghosh S, Agarwal A, Mishra MK and Sen E (2007) Kaempferol induces apoptosis in glioblastoma cells through oxidative stress. *Molecular Cancer Therapy* 6:2544–2553.

Somerset SM and Johannot L (2008) Dietary flavonoid sources in Australian adults. *Nutrition and Cancer* 60:442–449.

Stiborova M, Poljakova J, Mrizova I, Borek-Dohalska L, Eckschlager T, Adam V, Kizek R and Frei E (2014) Expression Levels of Enzymes Metabolizing an Anticancer Drug Ellipticine Determined by Electromigration Assays Influence its Cytotoxicity to Cancer Cells - A Comparative Study. *International Journal of Electrochemical Science* 9:5675-5689.

Termentzi A and Kokkalou E (2008) LC-DAD-MS (ESI+) Analysis and Antioxidant Capacity of Crocus sativus Petal Extracts. *Planta Medica* 74:573-581.

van der Woude H, Boersma MG, Alink GM, Vervoort J and Rietjens IMCM (2006) Consequences of quercetin methylation for its covalent glutathione and DNA adduct formation. *Chemico-Biological Interactions* 160:193-203.

Vazhayil BK, Sundaram SD, Annapandian VM, Abhirama BR, Sudha M, Thiyagarajan T, Kalpana KG and Pushpa S (2014) Natural Products and its Derived Drugs for the Treatment of Neurodegenerative Disorders: Alzheimer's Disease-A Review. *British Biomedical Bulletin* 2:359-370.

Verrastro I, Pasha S, Jensen KT, Pitt AR and Spicket CM (2015) Mass Spectrometry-Based Methods for Identifying Oxidized Proteins in Disease: Advances and Challenges. *Biomolecules* 5:378-411.

Walle T, Vincent TS and Walle UK (2003) Evidence of covalent binding of the dietary flavonoid quercetin to DNA and protein in human intestinal and hepatic cells. *Biochemical Pharmacology* 65:1603–1610.

Williams CA and Grayer RJ (2004) Anthocyanins and other flavonoids. *Natural Product Reports* 21:539-573.

Winkel-Shirley B (2001) Flavonoid biosynthesis. A colorful model for genetics, biochemistry, cell biology, and biotechnology. *Plant Physiology* 126:485-493.

Winkel-Shirley B (2002) Biosynthesis of flavonoids and effects of stress. *Current Opinion in Plant Biology* 5:218-223.

Wojdylo A, Oszmianski J and Czemerys R (2007) Antioxidant activity and phenolic compounds in 32 selected herbs. *Food Chemistry* 105:940–949.

Wojtunik-Kulesza KA, Oniszczuk A, Oniszczuk T and Waksmundzka-Hajnosa M (2016) The influence of common free radicals and antioxidants on development of Alzheimer's Disease. *Biomedicine & Pharmacotherapy* 78:39–49.

World Cancer Research Fund/American Institute for Cancer Research (2007) Food, Nutrition, Physical Activity, and the Prevention of Cancer: a Global Perspective. Washington DC: AICR.

Zeka K, Ruparelia KC, Continenza MA, Stagos D, Vegliò F and Arroo RR (2015) Petals of Crocus sativus L. as a potential source of the antioxidants crocin and kaempferol. *Fitoterapia* 107:128-134.

Zhang AL, Story DF, Lin V, Vitetta L and Xue CC (2008) A population survey on the use of common medicinal herbs in Australia. *Pharmacoepidemiology and Drug Safety* 17:1006–1013.

Editors' Contact Information

Dr. Teresa Garde-Cerdán,
Section of Viticulture and Enology,
Service Research and Technological DevelopmentAgrifood (CIDA)
Science Institute of Vine and Wine (ICVV)
Logroño, Spain
Email: teresa.gardecerdan@gmail.com

Dr. Ana Gonzalo-Diago,
Researcher,
Science Institute of Vine and Wine (ICVV),
Logroño, Spain
Email: anagonzalodiago@hotmail.com

INDEX

A

ABA, 20, 39, 49, 53
accessions, 16, 173
accounting, 47
acetaminophen, 130, 201, 213
acetylation, 177
acid, 2, 20, 22, 23, 29, 30, 31, 35, 39, 40, 42, 47, 53, 59, 63, 64, 65, 68, 69, 119, 121, 127, 135, 182, 205, 207
active compound, 102, 140
acute lung injury, 90
adenocarcinoma, 75, 76, 185
adenoma, 209
adhesion, 92
adiponectin, 84
adults, 83, 84, 213
adverse effects, 73, 200
Afghanistan, 181
Africa, 152
age, 22, 51, 85, 185
aggressiveness, 81
agriculture, 28, 30, 35, 36, 56
air pollutants, 206
airway epithelial cells, 131
airways, 86
alfalfa, 38
algae, 36
alkalinity, 31
alkaloids, 34, 35, 59, 199
allergic asthma, 90
allergic inflammation, 86, 93
allergic rhinitis, 96
allergy, 45
aloe, 129
alternative medicine, 123, 125, 127, 198, 210
alters, 51
aluminium, 42
amino, 20, 36, 40, 58, 173, 182, 188
amino acid(s), 40, 58, 173, 182, 188
ammonia, 3, 20, 30, 37, 38, 40, 59, 64
ammonium, 188
analgesic, 115, 185
angiogenesis, ix, 72, 77, 81, 92, 95, 102, 116
angiotensin converting enzyme, 139
animal food
Animal Food, 184
ANS, 4
anthocyanin(s), 2, 3, 14, 15, 16, 23, 26, 27, 34, 35, 39, 40, 43, 44, 45, 46, 47, 49, 50, 52, 53, 55, 56, 58, 59, 61, 65, 66, 142, 152, 172, 180, 187, 189, 193, 201, 205, 211, 212, 214
antibiotic, 67, 85
antibody, 201, 209
anti-cancer, 102, 184, 209
anticancer chemotherapeutics, 74
anticancer drug, 80, 116
anticonvulsant, 200

antidepressant, 191
anti-gas, 115
antigen, 75, 85
antioxidant, vii, 2, 10, 14, 17, 28, 34, 35, 36, 37, 39, 42, 45, 51, 53, 56, 57, 61, 62, 65, 66, 67, 69, 74, 82, 84, 85, 86, 93, 96, 115, 116, 117, 119, 122, 123, 124, 125, 129, 132, 136, 139, 140, 142, 143, 144, 147, 149, 150, 184, 185, 193, 197, 199, 201, 207, 208,209, 210, 212
antioxidative activity, 116, 135
antitumor, viii, 73, 74, 76, 77, 83, 87, 131
anxiety, 21
apex, 180
apoptosis, viii, 60, 72, 76, 77, 78, 79, 80, 81, 83, 86, 90, 91, 92, 93, 94, 95, 97, 98, 99, 102, 116, 119, 129, 130, 132, 133, 134, 149, 185, 191, 208, 211, 213
apples, 23, 34, 39, 54, 64, 202, 203
aqueous solutions, 204
Arabidopsis thaliana, 64, 106, 126, 145
Argentina, 181
aromatic rings, 2, 202
arrest, viii, 72, 77, 79, 94, 98
ARS, 136
arteriosclerosis, vii, 185
arthritis, 45, 207
ascorbic acid, 50
Asia, 148, 152, 183
asparagales, 164
assessment, 137
asthma, 85, 185
atherosclerosis, 83, 93, 136
ATP, 10, 79, 80, 97, 98
attitudes, 212
authentication, 15
automation, 199
auxins, 180
Azerbaijan, 181

B

bacteria, 24, 144
Bangladesh, 101
barriers, 48
base, 9, 179
Belgium, 182
beneficial effect, 28
benefits, viii, 22, 27, 28, 75, 88, 102, 186, 191, 198, 206
benzene, 43
benzodiazepine, 145
beverages, 17, 27, 102, 132, 197, 208
bile, 138
bioavailability, viii, ix, 72, 88, 89, 102, 103, 116, 207, 209
biochemistry, 57, 64, 117, 126, 195, 214
biodiversity, 121, 146
biological activities, 10, 125, 129, 134, 185, 200
biological activity, 22, 126, 136
biological control, 29
biomarkers, 83, 84, 210
biomass, 66
biosensors, 56
biosynthesis, v, vii, ix, 1, 2, 4, 11, 12, 13, 14, 16, 17, 21, 22, 26, 31, 37, 38, 39, 40, 42, 43, 45, 49, 51, 54, 58, 59, 60, 61, 68, 72, 124, 151, 153, 173, 174, 175, 176, 177, 195, 202, 203, 204, 212, 214
biosynthetic pathways, 46, 173, 202
biotechnology, 117, 190, 195, 214
biotic, vii, 2, 19, 24, 25, 26, 31, 45, 56
biotic factor, 26
black tea, 74
bladder cancer, 80, 90, 98
blood, 50, 77, 81, 82, 84, 85, 88, 115, 145, 185, 188, 200, 201, 207
blood pressure, 50, 145, 188, 200
blood vessels, 77, 81, 82
body weight, 83, 84, 85, 86
bonds, 205
brain, 115, 185
breast cancer, 51, 72, 79, 87, 90, 92, 127, 130, 145, 210
breast carcinoma, 94
breeding, 21, 35, 68, 186
buns, 182
buttons, 2, 43
by-products, 194, 211

C

c. sativus, 152, 153, 160, 164, 165, 167, 168, 169, 170, 177, 179, 180, 181, 183, 184, 201, 204
cabbage, 74, 75, 138
cancer, vii, viii, 21, 28, 67, 71, 72, 73, 74, 75, 76, 77, 78, 79, 81, 82, 86, 88, 90, 91, 93, 94, 95, 96, 97, 98, 102, 114, 115, 119, 121, 125, 130, 137, 138, 184, 187, 195, 197, 199, 208, 209, 211, 213
cancer cells, 67, 74, 76, 94, 130, 184
cancer death, 73, 77
cancer therapy, 95, 115, 187, 208
cancerous cells, 207
candidates, 74
capillary, 117, 134, 144
capsule, 178, 179
carbohydrate(s), 36, 52, 184
carbon, 43, 202
carcinogenesis, 76, 77, 80, 81, 82
carcinoma, 76, 86, 87, 93, 96, 132, 185
cardiovascular disease(s), 21, 74, 83, 102, 115, 136, 185, 209
cardiovascular disorders, vii
carotene, 201
carotenoids, 91, 125, 137, 182, 188
carotid arteries, 127
cascades, 83
case study, 115
CBS, 67
C-C, 120, 121, 133
cell biology, 195, 214
cell culture, vii, 19, 22, 29, 38, 52, 61, 62, 65, 66, 204, 212
cell cycle, 72, 77, 79, 94
cell death, 79, 115, 128, 184
cell killing, 98
cell line(s), 22, 72, 80, 81, 83, 89, 95, 97, 121, 146, 184, 185, 208
cell surface, 80, 92
cellulose, 31
cervical cancer, 79
challenges, 67, 91
chemical characteristics, 182
chemical reactions, 204
chemical structures, 73
chemical(s), 4, 6, 21, 25, 28, 29, 43, 52, 57, 58, 67, 68, 73, 88, 121, 124, 127, 131, 134, 138, 143, 147, 149, 172, 173, 182, 199, 201, 204
chemoprevention, 73, 74, 86, 90, 94, 187, 195, 208
chemotaxonomical value, 171, 172
chemotherapeutic agent, viii, 71, 73, 87
chemotherapy, viii, 71, 72, 73, 75, 87, 102, 116, 199
chemotypes, 172, 173
China, 101, 120, 121, 123, 124, 125, 127, 132, 133, 134, 138, 141, 143, 144, 145, 146, 147, 148, 149, 152, 181, 186
chinese medicine, 127
chitin, 31, 36, 38, 52, 53, 128
chitinase, 23
chitosan, 20, 30, 31, 36, 37, 42, 48, 49, 52, 53, 54, 56, 57, 58, 60, 61, 62, 63, 64, 66, 68
chlorophyll, 137
chloroplast, 189
cholesterol, 60, 83, 84, 93, 185
chromatography, 190, 194, 205
chromosome, 189
chronic diseases, 102
chronic fatigue, 207
circadian rhythm, 25
circulation, 88
classes, 15, 17, 29, 41, 43, 56
classification, 31, 162, 171, 172, 195
cleavage, 9, 16
climate, 16
clinical trials, 116, 208
closure, 174
C-N, 146
CO2, 9
coenzyme, 53
colic, 185
colitis, 96
collaboration, 194
colon, viii, 71, 72, 80, 81, 86, 87, 92, 94, 95, 99, 115, 133, 185, 207

colon cancer, 72, 80, 81, 87, 92, 94, 95, 99, 133
color, 3, 26, 49, 53, 182, 187, 190, 211, 212
colorectal cancer, 72, 86
colour stability, 63
commercial, 22, 30, 46, 64, 69, 182, 183
community, 199
complexity, 62
complications, 115
composition, vii, 14, 15, 16, 17, 19, 21, 22, 28, 29, 34, 42, 47, 50, 52, 54, 55, 59, 65, 66, 68, 123, 127, 136, 138, 140, 147, 154, 183, 192, 193
compost, 201
compounds, vii, viii, 2, 4, 6, 12, 14, 15, 16, 17, 18, 19, 20, 21, 22, 24, 25, 26, 27, 28, 30, 34, 37, 38, 42, 43, 44, 47, 49, 50, 51, 52, 60, 61, 63, 64, 65, 66, 69, 72, 73, 85, 87, 88, 90, 116, 118, 120, 121, 122, 124, 128, 130, 132, 134, 136, 139, 140, 141, 142,145, 146, 148, 149, 174, 182, 188, 191, 194, 199, 200, 201, 202, 204, 205, 207, 214
concise, 213
condensation, 174, 202
conference, 194
configuration, 182
conifer, 58
conjugation, 192
constituents, 13, 102, 117, 118, 119, 120, 121, 122, 123, 124, 125, 126, 127, 129, 130, 131, 134, 135, 137, 138, 140, 141, 142, 143, 144, 145, 146, 147, 148, 149, 150, 185, 187, 190, 194, 195, 204, 205, 213
consumers, 22, 29, 58
consumption, viii, 21, 71, 75, 84, 97, 114, 185, 190, 207, 213
contamination, 199
content, ix, 2, 11, 16, 17, 23, 26, 31, 34, 36, 37, 39, 40, 42, 45, 46, 47, 48, 49, 50, 52, 53, 54, 55, 61, 63, 64, 65, 66, 67, 68, 75, 89, 117, 123, 130, 136, 139, 145, 148, 150, 151, 152, 165, 167, 168, 170, 172, 184, 201, 202, 205, 209
control group, 49, 77, 84
controversial, 21
convergence, 56, 173
cooking, 59
cooling, 185
copper, 10
corm, 152, 178, 179, 193
coronary heart disease, 60
correlations, 14
cosmetic(s), 36, 60, 119, 197, 199, 201, 208
cost, 35, 36, 73, 197
co-therapy, viii, 74, 87
coumarins, 28
crises, 28
crocins, 182, 183, 192, 200, 201, 204
crocus, v, vi, viii, ix, 102, 103, 110, 132, 137, 138, 151, 152, 153, 154, 162, 163, 164, 165, 166, 168, 169, 170, 171, 172, 173, 174, 176, 177, 178, 179, 180, 181, 183, 184, 186, 187, 188, 189, 190, 191, 192, 193, 194, 195, 197, 198, 199, 200, 201, 208, 209, 210, 212, 213, 214, 215
crocus sativus, vi, viii, ix, 102, 103, 110, 137, 151, 152, 153, 187, 188, 189, 190, 191, 192, 193, 194, 195, 197, 199, 200, 201, 208, 209, 210, 212, 213, 214, 215
crocus species, ix, 151, 152, 153, 154, 163, 164, 165, 166, 169, 171, 172, 177, 180, 183, 186, 192
crop production, 28
crop(s), viii, 19, 21, 22, 24, 26, 28, 29, 30, 31, 34, 35, 36, 40, 46, 50, 51, 57, 58, 63, 65, 67, 181, 183, 186
CRP, 83, 84, 119
crude herb, 198
cultivars, vii, 11, 13, 15, 16, 18, 19, 29, 66, 118, 122, 131, 141, 168, 172, 192, 194
cultivated saffron, ix, 151, 152, 179
culture, 22, 26, 31, 69, 92, 130, 147, 183, 202, 213
culture conditions, 147
culture medium, 31
cure, 96
cuticle, 36
cyclins, 79

cyclodextrins, 43
cyclooxygenase, 99, 145
cytochrome(s), 85, 88, 207, 210
cytokines, 82, 83, 116
cytoplasm, 85, 87, 204
cytotoxicity, 143

D

damages, 86
danger, 29, 53
DART, 66
deacetylation, 36
deaths, 73
decay, 39
defence, 51, 63, 180
defense mechanisms, 25, 40, 60, 63
deficiencies, 25
deficiency, 80
deficit, 21
degradation, 86, 205, 206, 207
dehiscence, 179
dehydration, 23, 53, 182
depression, 184, 185, 186, 191, 192
derivatives, vii, 1, 2, 4, 6, 7, 9, 10, 11, 12, 35, 45, 63, 74, 88, 121, 139, 146, 150, 153, 164, 180, 199, 211
detection, 67, 136, 145, 205
detoxification, 207
DHS, 140
diabetes, 21, 84, 120, 185
diabetic patients, 84, 91
diet, viii, 12, 42, 62, 71, 74, 83, 84, 91, 197, 199, 202, 206, 207, 212
dietary fiber, 184
dietary intake, 20, 65, 210
digestibility, 184
dihydrokaempferol-3-rhamnoside, 4
dihydrokaempferol-rhamnoside, 8
dihydroquercetin-3-galactoside, 4
dihydroquercetin-3-glucoside, 4
dihydroquercetin-3-rhamnoside, 4
diseases, vii, viii, 19, 21, 22, 24, 28, 31, 35, 39, 43, 50, 51, 52, 61, 65, 67, 73, 76, 82, 83, 102, 103, 197, 199, 200, 206
dissociation, 14
distribution, 51, 134, 148, 154, 172, 181, 189, 194
diversity, 77, 152
DNA, 72, 77, 79, 80, 82, 90, 122, 187, 207, 212, 214
DNA damage, 79, 90
dosage, 116
double bonds, 206
down-regulation, 128
drought, 25, 39, 186
drug discovery, 128, 213
drug efflux, 97
drugs, 36, 73, 85, 87, 88, 102, 116, 184, 199, 212
dry matter, 200
drying, 183, 204, 205, 211

E

Easter, 124
ECM, 82
ecology, 57, 137, 143
egg, 36
Egypt, 118, 126
electric field, 65
electrons, 28, 206
electrophoresis, 117, 134, 144
elongation, 180, 194
elucidation, 139, 145, 147
encapsulation, viii, 72
endothelial cells, 60, 185, 190
energy, 9, 184
engineering, 21, 55, 61, 65
England, 182
enlargement, 185
environment, 20, 23, 29, 30, 204
environmental factors, 11, 12, 26, 30, 39, 45, 62
environmental impact, 14
environmental stress(es), 21, 29
enzyme(s), 11, 20, 24, 28, 31, 35, 37, 38, 39, 41, 44, 47, 51, 53, 57, 68, 69, 82, 85, 86, 93, 128, 132, 145, 174, 177, 185, 194, 202, 204, 207, 211

eosinophilia, 85
eosinophils, 85
epidemiological investigations, 102
epidemiology, 88, 138
epidermis, 2, 43
epithelial ovarian cancer, 56, 91, 125
epithelium, 76, 86
equipment, 205
ESI, 13, 17, 65, 124, 126, 139, 188, 193, 201, 209, 214
ester, 30, 35, 121
Estonia, 182
estrogen, 90, 135
ethanol, 34, 51, 52, 85, 205
ethers, 202
ethyl acetate, 146
ethylene, 23, 42, 48, 50, 52, 53, 57
Europe, ix, 151, 181, 186
European Union, 192
evidence, 4, 11, 55, 76, 177, 185, 194
evolution, 24, 189
excretion, 210, 213
expectorant, 200
exploitation, 22
exposure, 12, 22, 25, 42, 48, 76
external growth, 178
extracellular matrix, 82, 185
extraction, 21, 143, 167, 204, 205, 208, 211, 212
extracts, 6, 14, 38, 48, 49, 60, 122, 125, 128, 130, 132, 133, 136, 138, 141, 164, 182, 185, 187, 189, 193, 198, 200, 211
exudate, 130, 146

F

factories, 62
families, 202
farmers, 22, 29, 184
farms, 55
FAS, 80
fasting, 84
fatty acids, 31, 87
fermentation, 15
fertilization, 179
fiber(s), 36, 60, 184
fibroblasts, 83, 98
field trials, 40
films, 36
filtration, 205
financial, 51
Finland, 202
fitness, 27
flavonoids, viii, ix, 4, 9, 14, 15, 17, 23, 25, 26, 28, 34, 37, 38, 40, 43, 45, 46, 48, 50, 51, 57, 65, 68, 74, 76, 87, 88, 90, 91, 96, 97, 102, 117, 119, 120, 123, 124, 125, 127, 128, 129, 130, 131, 132, 133, 135, 137, 138, 139, 140, 141, 142, 143, 144, 145, 146, 147, 148, 149, 151, 152, 154, 164, 167, 168, 170, 172, 173, 174, 180, 189, 195, 201, 202, 204, 207, 209, 210, 213, 214
flavonol(s), vii, 1, 2, 3, 4, 5, 6, 10, 11, 12, 13, 14, 15, 16, 17, 19, 20, 21, 23, 24, 26, 38, 43, 44, 45, 46, 47, 48, 49, 50, 51, 54, 55, 56, 60, 61, 62, 65, 67, 71, 72, 75, 76, 81, 82, 83, 85, 86, 87, 89, 92, 93, 96, 98, 99, 116, 118, 120, 121, 122, 123, 124, 125, 126, 128, 129, 131, 134, 137, 138, 139, 141, 143, 145, 147, 153, 164, 174, 175, 177, 180, 181, 185, 187, 190, 191, 194, 197, 202, 203, 205, 206, 207, 209, 210, 211
flavor, 208
flavour, 204
floral bio-residues, 167, 184, 185, 193
flowers, ix, 26, 27, 62, 74, 95, 121, 123, 124, 126, 129, 137, 138, 140, 142, 143, 151, 152, 172, 178, 179, 180, 183, 184, 185, 186, 187, 192, 193, 194, 199, 200, 201
fluid, 85, 211
fluid extract, 211
fluorescence, 15
fluoxetine, 192
folate, 84
folic acid, 50, 53
food, vii, ix, 1, 21, 24, 27, 28, 36, 55, 75, 84, 88, 98, 114, 117, 118, 119, 120, 122,

123, 124, 127, 131, 132, 135, 139, 141, 142, 143, 145, 147, 151, 152, 153, 181, 183, 193, 197, 199, 202, 204, 205, 207, 208, 209, 211
food additive(s), ix, 27, 117, 151, 152, 199
food industry, 181
food products, 28, 183, 207
Ford, 68
formation, 9, 11, 47, 52, 77, 83, 174, 177, 180, 194, 201, 202, 207, 214
fragments, 6, 9
France, 181, 182, 202
free kaempferol, 164
free radicals, 10, 28, 206, 214
fruits, vii, viii, 2, 17, 23, 26, 27, 30, 31, 34, 38, 39, 40, 42, 46, 47, 48, 50, 51, 52, 54, 60, 61, 68, 71, 74, 75, 88, 97, 102, 103, 116, 122, 131, 132, 135, 141, 144, 145, 188, 197, 199, 206, 212
function, 25, 31, 56, 79, 94, 95, 99, 116, 127, 132, 153, 177, 180, 191
functional food, 24, 30, 197, 208
fungal infection, 38
fungi, 11, 24
fungus, 40, 63

G

gamma radiation, 30
gene expression, 23, 45, 52, 57, 58, 61, 79, 80, 81, 131
genes, 13, 14, 21, 29, 35, 45, 46, 49, 51, 52, 56, 59, 60, 68, 80, 98, 124, 186
genetic background, 18
genetic engineering, 21
genetic information, 174
genetics, 29, 195, 214
genomic instability, 77
genotype, 2, 23, 176
genus, ix, 146, 151, 152, 153, 164, 173, 181, 191, 192
geographical origin, 61, 168, 188
Germany, 186, 193, 194
germination, 39, 42, 50, 55, 59, 61, 66, 152, 177
ginger, 74
gingival, 200
glioblastoma, 76, 83, 97, 133, 208, 213
glioma, 76, 87, 128
global scale, 198
glucose, 2, 23, 36, 84, 123, 141, 174, 182, 191
glucoside, 4, 7, 9, 12, 48, 50, 98, 118, 135, 142, 163, 164, 165, 167, 168, 169, 170, 172, 174, 176, 177
glucosinolates, 27, 52, 119
glutamate, 83, 99
glutamic acid, 50
glutathione, 84, 86, 214
glycine, 39, 59
glycogen, 23, 115
glycoproteins, 31
glycoside, vii, viii, 1, 9, 10, 17, 102, 119, 121, 123, 124, 126, 131, 134, 137, 142, 143, 165, 167, 168, 169, 172, 174
glycosylated hemoglobin, 84
glycosylation, ix, 13, 14, 152, 153, 172, 174
glycosyltransferases, 174
gold nanoparticles, 97
gracilis, 112, 118
grazing, 184
Greece, 181, 202
greenhouse(s), 23, 48, 66
grouping, 172, 200
growth, 11, 22, 25, 30, 34, 54, 66, 76, 77, 78, 79, 82, 91, 93, 179, 185, 204
growth factor, 91, 93
growth rate, 82

H

habitat, 172, 204
harvesting, 47, 181, 184
HDAC, 87
healing, 45, 123
health, vii, viii, 24, 28, 29, 30, 44, 55, 57, 60, 65, 69, 71, 86, 102, 142, 191, 198, 199, 206, 207, 213
health care, 198
health problems, 207

heart disease, 207
helicobacter pylori, 82, 129
hematology, 209
heme, 4
hemorrhoids, 45
hepatitis, 82, 132
hepatocytes, 85, 213
hepatoma, 81, 95, 137
hepatotoxicity, 85, 130, 201
herbal medicine, vii, 210
high fat, 84
highlands, 51
high-resolution mass spectrometry, vii, 1, 15, 17
histology, 201, 209
histone, 72, 80, 89
histone deacetylase, 72, 80, 89
history, 181
HIV-1, 118
homeostasis, 81, 191
homolytic, 9
homovanillic acid, 92, 211
Hong Kong, 101
hormone(s), 39, 58, 180
horticultural crops, 67
host, 26, 29, 51, 52
human, vii, 12, 19, 20, 21, 26, 28, 43, 50, 58, 60, 63, 67, 72, 74, 77, 80, 82, 86, 87, 89, 90, 92, 93, 94, 95, 96, 97, 98, 99, 120, 121, 128, 131, 132, 133, 134, 136, 138, 185, 187, 190, 198, 202, 204, 206, 207, 208, 209, 214
human body, 206
human health, viii, 19, 20, 21, 26, 28, 43, 50, 58, 63, 90
human immunodeficiency virus (HIV), 118, 136
human neutrophils, 120, 136
hybrid, 14
hybridization, 189
hydrazine, 96
hydrogen, 60
hydrogen peroxide, 60
hydrolysis, 204, 205
hydroxyl, 2, 28, 177, 202, 206, 212
hydroxyl groups, 2, 28, 202, 206
hypoxia, 81, 99
hypoxia-inducible factor, 81, 99

I

ideal, 73, 74
identification, ix, 6, 7, 9, 12, 17, 22, 117, 132, 136, 139, 144, 149, 162, 188
IFN, 123
IL-13, 85
IL-8, 83
ileum, 188
immune defense, 29
immune system, 24, 29, 201
immunity, 35, 58
immunoglobulin, 93
improvements, 199
in vitro, 21, 22, 26, 38, 50, 61, 72, 74, 76, 77, 79, 83, 85, 86, 87, 88, 95, 98, 115, 120, 125, 127, 128, 129, 139, 149, 185, 212
in vivo, viii, 50, 72, 74, 76, 77, 83, 88, 89, 98, 115, 119, 127, 139, 149, 200
incidence, viii, 56, 64, 71, 73, 74, 91, 96, 125, 199
incongruity, 88
India, 67, 181
individuals, 21
Indonesia, 209
inducer, 30, 35, 52, 77, 87
induction, viii, 22, 26, 30, 35, 38, 42, 46, 63, 64, 66, 72, 77, 78, 80, 93, 127, 132, 208
industries, 36, 199
industry, 22, 27, 182, 194, 197, 198, 199, 207
infection, 24, 29, 42, 53, 60, 64
inflammation, ix, 45, 72, 78, 82, 83, 84, 86, 90, 91, 96, 102, 116, 120, 185
inflammatory cells, 85
inflammatory disease, 82
inflammatory mediators, 123
inflammatory responses, 72, 92
influenza, 128
ingestion, 76

ingredients, 75, 131, 204, 205, 210, 211
inhibition, 50, 72, 75, 77, 81, 83, 85, 86, 88, 90, 95, 99, 128, 180, 190, 213
inhibitor, 77, 79, 87, 89
initiation, 39, 73
injury, 42, 98
insects, 24, 26, 63, 68
insulin, 84, 85, 94
insulin resistance, 94
integrity, 52
internalization, viii, 72
intervention, 76, 83, 84
ion channels, 31
ionization, 6, 13, 16, 18, 148
ions, 9, 28
Iran, 181, 184, 189, 194, 195
Ireland, 54
iridaceae, ix, 110, 151, 164, 188, 189, 190, 191, 192, 194, 195, 199
irradiation, 37, 48
irrigation, 12, 21
IRS, 85
ischemia, 193
Islam, 193
islands, 182
isoflavonoids, 147
isolation, ix, 103, 122, 139, 144, 190, 193
isorhamnetin, 2, 4, 43, 45, 153, 172
isorhamnetin rutinoside, 4
Israel, 191
Italy, 1, 35, 181, 182, 183, 187, 197, 202

J

Japan, 181, 190
Jordan, 194

K

kaempferide-*p*-coumaroylhexoside, 4, 7, 10
kaempferol 3-*O*-sophoroside, 163, 167, 168, 185
kaempferol 3-*O*-β-sophoroside, 163, 168
kaempferol acylated glycosides, 164, 170, 171
kaempferol *C*-glycosides, 170
kaempferol chemistry, 12
kaempferol composition, 154, 155, 156, 157, 158, 159, 160, 161, 162
kaempferol diglycosides, 165, 166
kaempferol glycosides, ix, 103, 115, 127, 132, 137, 151, 152, 153, 162, 163, 164, 171, 174, 176, 177, 180, 182, 208
kaempferol monoglycosides, 164, 165
kaempferol rutinoside, 4, 12
kaempferol triglycosides, 168, 169
kaempferol-3-*O*-glucoside, 12
kaempferol-rutinoside, 8, 12
karyotype, 189
kidney, 42, 50, 55, 60, 85
kill, 73
kinase activity, 79, 90, 93
Korea, 101, 133

L

landscape, 198
laricitrin, 2, 3, 45
latency, 180
Latvia, 194
LC-MS, 141, 205, 212
LC-MS/MS, 141
LDL, 10, 15, 83, 84, 97
learning, 200
leaves, ix, 23, 26, 34, 36, 38, 47, 48, 57, 59, 60, 62, 66, 74, 90, 117, 118, 120, 121, 122, 123, 125, 126, 127, 128, 129, 130, 131, 132, 133, 134, 135, 138, 140, 141, 142, 143, 145, 146, 147, 148, 149, 150, 151, 153, 156, 158, 160, 161, 162, 170, 178, 179, 184,194, 195
Lebanon, 183
legislation, 22
legume, 29, 50, 61, 84, 91
leukemia, 72, 76, 86, 96, 185, 208
liberty, 53
life-cycle, 153, 177, 178, 179
lifetime, 52

ligand, 92
light, 2, 11, 13, 25, 26, 42, 45, 51, 66, 138, 204
light conditions, 13, 51
linoleic acid, 201
lipid oxidation, 213
lipid peroxidation, 11, 96
lipids, 38, 85, 184
liquid chromatography, 14, 15, 16, 17, 18, 67, 136, 143, 148, 189
liver, 98, 115, 138, 185, 201
loci, 189
low temperatures, 30
lung cancer, 76, 90, 114, 121, 185
Luo, 60, 74, 79, 81, 85, 87, 88, 93, 94, 95, 104, 106, 115, 133, 134, 146, 147, 184, 191
lutein, 84, 189
lymph, 115

M

machinery, 22
macromolecules, 201, 206
macrophages, 86, 123
magnitude, 15
majority, 152
Malaysia, 51, 125
management, 12, 29, 52, 58, 65
manipulation, 21, 51
MAPK/ERK, 79
marketing, 30
mass, vii, 1, 6, 7, 13, 15, 16, 17, 18, 66, 67, 136, 145, 148, 189
mass spectrometry, vii, 1, 6, 7, 13, 15, 16, 17, 18, 66, 67, 136, 145, 148, 189
mastitis, 86, 89
materials, 22, 127, 133, 134, 135, 140, 143, 146, 147, 148, 149, 150
matrix, 6, 82, 94
matrix metalloproteinase(s) (MMPs), 82, 94, 133
maturation process, 47
MCP, 20, 23, 52
media, 22, 189
medical, 64, 198, 200, 201
medical care, 198
medicine, vii, 19, 20, 36, 43, 44, 73, 90, 117, 119, 138, 198, 200, 211, 213
Mediterranean, 182
Mediterranean countries, 182
medulloblastoma, 93
melts, 178
membranes, 204
menstruation, 185
metabolic, 212
metabolic pathways, 22
metabolism, 12, 21, 34, 36, 37, 38, 42, 50, 52, 58, 64, 85, 88, 136, 206, 210, 213
metabolites, vii, 1, 6, 11, 15, 19, 20, 21, 22, 24, 25, 26, 27, 28, 29, 31, 33, 34, 37, 38, 42, 43, 51, 52, 54, 55, 57, 59, 61, 62, 65, 67, 68, 69, 96, 119, 123, 125, 211
metabolized, 102, 116, 207
metal ion(s), 25, 31
metalloproteinase, 72
metals, 28
metastasis, ix, 73, 78, 81, 82, 90, 102, 116
methanol, 54, 133, 193
methodology, 173
methylation, 202, 214
Mexico, 181
mice, 77, 85, 86, 89, 90, 96, 98, 126, 130, 189
micronutrients, 80
microorganisms, 28, 207
microRNA, 64
Middle East, ix, 151
migration, 93, 133
mildew, 23, 35, 37, 40, 48, 58, 62, 63
Miocene, 138
mitochondria, 92
mitosis, 79
MMP-2, 82
MMP-9, 82
model system, 201
models, 76, 77, 84, 85, 88, 89, 124, 188, 206
modernization, 183
modifications, 15, 35

moisture, 180, 204
mold, 35, 37, 54, 58, 64
molecular weight, 6, 20, 29, 31, 36, 56, 103, 164, 174
molecules, 2, 14, 29, 30, 174, 188, 198, 202
mollusks, 24
Moon, 106, 137
morbidity, 73
Morocco, 181
morphogenesis, 180, 181, 187
mortality, 73, 81, 83
motif, 207
mRNA, 79, 81, 88
mucin, 86
mucosa, 86
multiple mass spectrometry, 6, 7
multiplication, 77
mung bean, 42
muscle relaxant, 137
mutation(s), 15, 77
myocardial infarction, 125, 136
myricetin, 2, 3, 4, 23, 43, 45, 47, 48, 49, 50, 54, 61, 62, 93, 153, 172, 206, 207, 212, 213
myricetin hexoside-glucuronide, 4

N

nanoparticles, 88, 95
nanopharmaceuticals, viii, 72
National Academy of Sciences, 92, 191
natural compound, 73
natural food, 200
natural resources, 103
nausea, 200
nerve, 200
Netherlands, 97, 186, 192
neurodegenerative diseases, 21
neurons, 83
neurotoxicity, 83
neurotransmitter, 83
New Zealand, 181
next generation, 91
nitric oxide, 99, 116, 118, 131
nitric oxide synthase, 118
NMR, 126
N-N, 16
non-cancerous cells, 76
non-enzymatic antioxidants, 84
normal development, 41
North Africa, ix, 151
Norway, 48, 61
Norway spruce, 48, 61
nuclear magnetic resonance, 16
nucleic acid, 185
nucleus, 85
nutraceutical, vii, 1, 12, 21, 42, 44, 54, 82, 88, 191, 197
nutrient(s), 22, 25, 26, 69, 75, 81, 178, 212
nutrition, 43, 65, 89, 125, 134
nutritional status, 23

O

occurrence, ix, 65, 151, 153, 164, 192, 197, 209
oesophageal, 96
oil, 27
oligomers, 135
oncogenesis, 73
opportunities, 91
optimization, 22
oral cavity, 72, 76
organelles, 204
organism, 129
organ(s), 21, 36, 73, 77, 181
ornamental plants, 186, 188
ornamental use, ix, 152, 153
osmotic pressure, 31
osmotic stress, 39
osteoporosis, 21
ovarian cancer, 72, 76, 79, 81, 94, 95, 115, 134, 191
overweight, 84
oxidation, 10, 85, 204, 206, 207
oxidative damage, 65, 84, 207, 209
oxidative stress, 89, 92, 97, 99, 129, 206, 211, 213
oxygen, 65, 81, 82, 201, 206
ozone, 42

P

p53, 79, 83, 92, 94, 95, 98, 133, 211
paclitaxel, 73, 184, 199
Pakistan, 51, 66, 117, 140
PAL, 3, 20, 22, 23, 31, 34, 36, 37, 38, 39, 40, 50
palliative, 75
palmate, 11
pancreas, 75, 77, 115
pancreatic cancer, viii, 71, 75, 76, 94, 96, 99, 114, 138, 149, 194
parallel, 179
patents, 128, 204
pathogens, 22, 25, 28, 30, 31, 36, 40, 46, 53, 63, 68
pathway(s), viii, ix, 3, 4, 14, 22, 30, 35, 37, 39, 40, 43, 44, 45, 46, 49, 50, 56, 59, 61, 66, 72, 78, 79, 80, 81, 82, 85, 86, 91, 92, 94, 95, 96, 98, 99, 102, 116, 133, 139, 152, 153, 173
pedicel, 178, 179
peptide, 97
perianth, ix, 151, 152
permeability, 86, 88, 116
peroxidation, 145, 212
pertussis, 185
pesticide, 63
pests, vii, 19, 22, 50
Petal, 210, 214
PGE, 127
pH, 187, 206, 207
pharmaceutical(s), 22, 27, 117, 120, 122, 126, 127, 132, 135, 136, 137, 140, 146, 185, 197, 198, 204, 205, 208
pharmacokinetics, ix, 89, 96, 102, 209
pharmacology, 36, 129, 135, 141, 210, 213
phenolic compounds, viii, 2, 15, 16, 18, 19, 20, 22, 24, 25, 26, 27, 28, 30, 34, 42, 43, 47, 50, 51, 61, 64, 65, 116, 118, 122, 124, 130, 132, 134, 146, 148, 201, 214
phenotype(s), 82
phenylalanine, 3, 4, 20, 22, 30, 37, 38, 40, 43, 59, 64, 173
phosphates, 40, 63
phosphorylation, 72, 80, 85, 92, 120, 136, 211
physical properties, 60
physiological mechanisms, 206
physiology, 133, 149, 208
phytochemicals, viii, 17, 23, 65, 71, 72, 73, 74
phytomedicine, 137
phytotherapy, 213
PI3K, 72, 79
PI3K/AKT, 79
picrocrocin, 182, 188
pigmentation, 187, 193
pilot study, 75
placebo, 75, 84, 91, 98, 191
plant disease(s), 22, 28, 30, 32, 55, 67
plant growth, 26, 31, 34, 37, 38, 39, 63, 180
plants, vii, viii, 1, 19, 20, 21, 23, 24, 25, 26, 27, 28, 29, 30, 31, 34, 35, 37, 38, 40, 42, 43, 44, 46, 49, 50, 51, 53, 54, 55, 58, 60, 62, 64, 65, 66, 102, 103, 115, 122, 138, 144, 147, 153, 174, 180, 186, 187, 195, 198, 199, 202, 203, 204, 205
plasma levels, 84
plasminogen, 82
plastid, 192
playing, 27
PM, 117, 194
polar, 6, 120
polarity, 6
pollen, 141, 153, 180, 190, 191
pollination, 180, 181
pollinators, 180
polyamines, 31
polymer, 36
polymerization, 205
polyp, 209
polyphenols, 6, 14, 15, 26, 28, 31, 37, 41, 43, 68, 131, 149
polysaccharides, 31
population, 75, 90, 121, 138, 198, 215
Portugal, 118, 181
post-hoc analysis, 84
potassium, 201
potato, 66

predator, 25
preparation, 60, 173
preservative, 36
prevention, viii, 21, 28, 71, 74, 83, 85, 103, 199, 208, 209
primary tumor, 77
priming, 54
principles, ix, 95, 144, 152, 153, 190
prodrugs, 209
producers, 181
professionals, 198
pro-inflammatory, 82, 83, 85, 86, 92, 130
proliferation, 60, 64, 77, 79, 81, 88, 90, 96, 99, 145, 146, 149
propagation, 21
prostate cancer, 75, 76, 89, 93, 97, 98, 188
protection, 22, 36, 53, 57, 61, 62, 63, 68, 83, 85, 180, 204, 209
protective mechanisms, 37
protective role, 12, 43, 82, 212
protein synthesis, 187
proteinase, 23
proteins, 10, 31, 38, 52, 56, 59, 68, 79, 80, 85, 120, 136
pruning, 21
PSA, 75, 98
PTEN, 80, 98
pulp, 65
pumps, 87
purification, ix, 205

Q

quality control, 183, 193
quantification, 17, 124, 131, 139, 143, 164, 189
quantitative estimation, 129
quercetin, 2, 4, 23, 43, 45, 47, 48, 49, 50, 61, 62, 87, 91, 93, 94, 96, 116, 123, 131, 146, 152, 172, 181, 191, 194, 197, 204, 205, 206, 207, 208, 210, 211, 212, 213, 214

R

radiation, 11, 25, 37, 38, 48, 55, 62, 66, 204, 206
radicals, 28, 201, 206
Ramadan, 111, 140
reactions, 21, 29, 40, 174
reactive oxygen, 10, 11, 20, 92, 180, 211
receptor(s), 29, 30, 31, 53, 80, 85, 86, 92, 98, 133, 135, 145
recognition, 14, 29, 53
recovery, 85, 86
rectum, 111, 137
recurrence, viii, 71, 209
red wine, 3, 13, 14, 15, 17, 18, 53, 56, 60, 61, 63
regeneration, 45
relatives, 190
relaxation, 143
renaissance, 53
repair, 37
repressor, 92
reproduction, 25, 27
requirement(s), 12, 95, 116, 198, 199
researchers, vii, 19, 23, 36, 39, 88, 102, 205
reserves, 180
residue(s), 9, 92, 165, 167, 168, 170, 183, 184, 185, 193, 211
resistance, viii, 20, 22, 28, 29, 30, 32, 35, 36, 40, 46, 52, 53, 54, 55, 56, 57, 58, 59, 60, 62, 63, 64, 65, 67, 68, 69, 71, 73, 77, 80, 81, 87, 92, 184
resolution, vii, 1, 6, 7, 9, 12, 15, 17
response, 15, 22, 26, 29, 42, 43, 53, 55, 68, 69, 85, 94, 98, 145, 201, 204, 209
restoration, 85
resveratrol, 38, 42, 57, 58, 61, 66
reticulum, 91
reverse transcriptase, 136
rheumatoid arthritis, 197
rhizome, 98, 136
Rhizopus, 36
rings, 43

risk, 21, 28, 51, 74, 75, 76, 88, 91, 93, 96, 97, 98, 102, 114, 125, 136, 138, 185, 194, 198, 207, 209
RNA, 187
robotics, 199
root(s), 62, 66, 68, 69, 122, 124, 128, 134, 178, 179, 180
rules, 204

S

safety, ix, 56, 102
saffron floral bio-residues, 165, 167, 168, 185
safranal, 182, 190, 204
salinity, 25
salts, 31
SAR, 20, 22, 26, 35, 40, 46, 52
scavengers, 138
schema, 80
science, 119, 136, 141, 143, 190
seasonality, 25
secondary metabolism, 55, 59, 68, 173
secrete, 87
sedative, 200
seed, 25, 39, 53, 60, 66, 152, 179
seeding, 66, 75
seedling development, 152
selectivity, viii, 21, 71, 76
senescence, 39, 180
sensing, 53, 120
sensitization, 85
Serbia, 144
serum, 136, 190
serum albumin, 190
services, 198
sexual reproduction, 152
shade, 37
sheep, 184
shoots, 34, 65, 178, 179
showing, 34, 72, 76, 185
side effects, 73, 198
signal transduction, viii, 35, 69, 102, 116
signaling pathway, 29, 85, 90, 94
signalling, 78, 80, 82
signals, 53
silver, 48, 66
skeletal muscle, 143
skin, 2, 11, 13, 43, 45, 49, 51, 59, 65, 79, 115, 185, 199
skin cancer, 79
smoking, 206
smooth muscle, 97
sodium, 86, 96, 127
solubility, 201
solution, viii, 72, 200
sources, v, vii, ix, 1, 12, 21, 27, 29, 42, 60, 64, 74, 101, 102, 103, 151, 164, 186, 200, 202, 203, 209, 211, 213
South Africa, 62
soybeans, 74
Spain, 19, 71, 91, 125, 151, 179, 181, 182, 184, 186, 187, 202
species, ix, 10, 11, 16, 20, 35, 36, 38, 49, 65, 82, 92, 103, 142, 146, 148, 151, 152, 153, 154, 162, 163, 164, 165, 166, 168, 169, 171, 172, 173, 174, 177, 179, 180, 181, 183, 186, 188, 189, 190, 192, 195, 200, 202, 206, 211
spectrophotometry, 6
spectroscopy, 126
spice, ix, 151, 152, 167, 170, 181, 182, 183, 185, 188, 190, 193, 198, 199, 200, 201, 208
spleen, 185, 201, 209
Spring, 117, 168, 170, 192
sprouting, 180
squamous cell, 75
squamous cell carcinoma, 75
stability, 3, 49, 190, 205, 211, 212
stamens, 183, 184, 201
starch, 183
state, 29
sterile, 152
sterols, 38
stigma(s), ix, 151, 152, 153, 160, 167, 168, 170, 177, 181, 183, 184, 185, 188, 189, 190, 192, 199, 200, 201, 204
stimulant, 200
stimulation, 37, 115, 188

stimulus, 120
storage, 30, 37, 47, 56, 61, 64, 68, 123, 211
strategy use, 19
stress, 4, 24, 25, 26, 30, 37, 38, 39, 42, 49, 54, 56, 58, 61, 68, 91, 195, 206, 214
stress factors, 24, 25, 26
stress response, 30, 39, 56
stroke, 136
structure, 2, 4, 28, 43, 50, 51, 56, 57, 88, 92, 95, 96, 118, 139, 145, 173, 200, 202, 207
styles, 183
substitutes, 2, 45
substitution, 2, 43
substrate, 22, 85, 88, 91, 174, 177
sulfate, 86, 96, 102, 116
Sun, 90, 98, 113, 114, 123, 132, 134, 142, 144, 146, 147, 149, 150
suppression, 67, 72, 77
surveillance, 76
survival, 83, 96
survivors, 210
susceptibility, 36
suspensions, 55
Sweden, 182
Switzerland, 181
symptoms, 45, 85, 86, 207
synergistic effect, 43, 83
synthesis, 2, 11, 22, 27, 29, 33, 34, 37, 41, 43, 45, 46, 49, 50, 65, 97, 115, 185
Syria, 183
syringetin, 2, 3, 4, 44, 45

T

tamoxifen, 88, 96
tannins, 27, 28, 59, 128
target, 79, 80, 81, 91, 93, 94, 97
taxa, 162, 164, 165, 167, 168, 170, 172, 173, 174
taxonomic assignment, 154, 162
taxonomy, 43, 189
techniques, 6, 12, 21, 88, 204
temperature, 11, 13, 25, 31, 45, 51, 62, 66, 138, 178, 205, 211
tepals, ix, 152, 153, 154, 155, 156, 157, 158, 159, 160, 161, 162, 164, 165, 167, 168, 169, 170, 180, 183, 187, 189, 192, 212
therapeutic effect(s), 87
therapeutic use, vii
therapeutics, 198
therapy, viii, 73, 74, 87, 91, 130
thermal degradation, 183
thinning, 21
tissue, 26, 73, 77, 212
TLR4, 86
TNF-α, 82, 83, 84, 85, 86, 127
tobacco, 38, 56, 59
toxic effect, 87
toxicity, 88, 102, 116, 124, 136, 185, 191, 200, 213
toxicology, 129
traditional practices, 198
traits, 58, 172
transaminases, 85
transcription, 79, 81, 82, 129
transcription factors, 79
transduction, 78, 79
transformation, 26, 79, 82, 93, 94, 96
translocation, 65
transplant, 211
transport, 92, 180
treatment, viii, 21, 23, 29, 34, 36, 38, 39, 40, 45, 46, 48, 51, 53, 54, 56, 60, 61, 66, 69, 71, 73, 74, 76, 79, 83, 86, 89, 91, 102, 103, 121, 173, 184, 185, 186, 191, 192
trial, 75, 83, 84, 89, 91, 94, 98, 191, 192, 209
triggers, 38
triploid, 152
tumor, viii, 71, 72, 73, 76, 77, 79, 80, 81, 82, 87, 88, 90, 92, 97, 98, 99, 145, 185
tumor cells, viii, 72, 73, 76, 77, 80, 82, 87, 97, 185
tumor development, viii, 72, 79
tumor growth, 73, 77, 90
tumor necrosis factor (TNF), 80, 82, 83, 84, 85, 86, 92, 127
tumor resistance, 80
tumorigenesis, 76

tumors, 75, 77, 80, 81, 87, 185
tumours, 207
Turkey, 136, 181, 188, 190, 195
type 2 diabetes, 76, 84, 89

U

ulcerative colitis, 86
United Nations, 188
United States (USA), 14, 15, 16, 17, 54, 58, 92, 96, 181, 191
urinary bladder, 28
urokinase, 82
USDA, 75, 89
uterus, 115
UV irradiation, 51, 64
UV radiation, 25, 31, 38

V

vacuole, 204
vacuum, 205
Valencia, 61, 125, 187
vanadium, 120
variations, 45
varieties, 2, 12, 13, 14, 16, 21, 28, 45, 48
vas deferens, 188
vascular endothelial growth factor (VEGF), 81, 95, 97
vasodilation, 17
vegetables, vii, viii, 27, 31, 37, 38, 40, 50, 52, 54, 68, 71, 74, 75, 83, 88, 102, 103, 197, 199, 206, 209, 212
vegetative reproduction, 191
VEGF expression, 81, 95
vein, 60
vertebrates, 24, 36
viral diseases, 35
virus infection, 82
viruses, 11, 24, 25
vision, viii, 19
vitamin C, 53, 124
vitamin E, 97
vitamins, 84, 136
vitis grape varieties, 12

W

Washington, 15, 58, 214
waste, ix, 139, 198, 201, 205, 208
water, 63, 127, 178, 182, 200, 205
wavelengths, 43
wealth, ix, 102
Western countries, 81
Western Europe, 152
World Health Organization (WHO), 73, 74, 98, 198
worldwide, 28, 73, 185
wound healing, 121

X

xanthones, 138, 150
xenografts, 77

Y

yeast, 38, 49, 54, 59, 63, 66
yield, 12, 28, 35, 38, 40, 54, 64, 167, 212
Yugoslavia, 202